CLIMATE CHANGE, CULTURE, AND ECONOMICS: ANTHROPOLOGICAL INVESTIGATIONS

RESEARCH IN ECONOMIC ANTHROPOLOGY

Series Editor: Donald C. Wood

Recent Volumes:

Volume 24: Markets and Market Liberalization: Ethnographic Reflections – Edited by N. Dannhaeuser and C. Werner

Volume 25: Choice in Economic Contexts: Ethnographic and Theoretical Enquiries – Edited by D. Wood

Volume 26: The Economics of Health and Wellness: Anthropological Perspectives – Edited by D. Wood

Volume 27: Dimension of Ritual Economy – Edited by P. McAnany and E. C. Wells

Volume 28: Hidden Hands in the Market: Ethnographies of Fair Trade, Ethical Consumption and Corporate Social Responsibility – Edited by Donald Wood, Jeffrey Pratt, Peter Luetchford, and Geert De Neve

Volume 29: Economic Development, Integration, and Morality in Asia and the Americas – Edited by Donald C. Wood

Volume 30: Economic Action in Theory and Practice: Anthropological Investigations – Edited by Donald C. Wood

Volume 31: The Economics of Religion: Anthropological Approaches – Edited by Lionel Obadia and Donald C. Wood

Volume 32: Political Economy, Neoliberalism, and the Prehistoric Economies of Latin America – Edited by Ty Matejowsky and Donald C. Wood

Volume 33: Engaging with Capitalism: Cases from Oceania – Edited by Fiona McCormack and Kate Barclay

Volume 34: Production, Consumption, Business and the Economy: Structural Ideals and Moral Realities – Edited by Donald C. Wood

RESEARCH IN ECONOMIC ANTHROPOLOGY VOLUME 35

CLIMATE CHANGE, CULTURE, AND ECONOMICS: ANTHROPOLOGICAL INVESTIGATIONS

EDITED BY

DONALD C. WOOD

Department of Medical Education,
Akita University Graduate School of Medicine,
Akita, Japan

Emerald

United Kingdom – North America – Japan
India – Malaysia – China

Emerald Group Publishing Limited
Howard House, Wagon Lane, Bingley BD16 1WA, UK

First edition 2015

British Library Cataloguing in Publication Data
A catalogue record for this book is available from the British Library

ISBN: 978-1-78560-361-7
ISSN: 0190-1281 (Series)

CONTENTS

LIST OF CONTRIBUTORS

Olivia Barnett-Naghshineh	Anthropology, School of Social Sciences, University of Auckland, New Zealand
Ana Maria de Souza Mello Bicalho	Departamento de Geografia, Universidade Federal do Rio de Janeiro, Rio de Janeiro, Brazil
Sophie Bouly de Lesdain	EDF Research & Development, Associate IIAC LAU Lab (CNRS-EHESS), Clamart, France
Jane I. Guyer	Department of Anthropology, Johns Hopkins University, Baltimore, MD, USA
Scott William Hoefle	Departamento de Geografia, Universidade Federal do Rio de Janeiro, Rio de Janeiro, Brazil
Hao Li	Changjiang River Scientific Research Institute, Changjiang River Water Commission, Wuhan, China
Ty Matejowsky	Department of Anthropology, University of Central Florida, Orlando, FL, USA
Patrick B. Patterson	Population Health and Inequities Research Centre, O'Brien Institute for Public Health, University of Calgary, Calgary, Alberta, Canada
Victoria C. Ramenzoni	Social Science Team, Office of Program Planning and Integration, National Oceanic and Atmospheric Administration, Department of Commerce, Silver Spring, Maryland, USA

Richard K. Reed	Department of Sociology and Anthropology, Trinity University, San Antonio, TX, USA
Pasang Yangjee Sherpa	Department of Anthropology, Pennsylvania State University, University Park, PA, USA
Monica L. Smith	Department of Anthropology, University of California at Los Angeles, CA, USA
Meine Pieter van Dijk	UNESCO-IHE Institute for Water Education, Delft, and Maastricht School of Management, Maastricht, Netherlands
Donald C. Wood	Department of Medical Education, Akita University Graduate School of Medicine, Akita, Japan

EDITORIAL ADVISORY BOARD

INTRODUCTION

It is always a pleasure to start an REA volume introduction with a report of a new development in the series' long and proud history – now at the 37 year mark. This time, in fact, there are several. First, as some readers may have already noticed, REA is now officially sponsored by the Society for Economic Anthropology (SEA), a section of the American Anthropological Association. Volume 34 of REA (published in September 2014) was the first to bear this distinction, and the present volume is the second. Solidifying the relationship between SEA and REA – which had actually existed in an informal sense for many years – is expected to bring many benefits to both parties. One direct benefit to SEA members is the opportunity to purchase REA volumes at a discounted rate for a set amount of time following their publication. Please see the SEA website, or the REA corner on Emerald's website, for more information.

Another development worth reporting is the establishment of REA's Editorial Advisory Board (EAB), which was finalized shortly after Volume 34's publication. This ten-member panel of noted researchers will help support REA in various ways, such as by soliciting papers by experts in their fields, by assisting with the peer-review process when needed, and by spreading the word about new volumes. They will also lend their advice on REA projects and help steer the development of the series in general. They are (in alphabetical order): Jeffrey Cohen (The Ohio State University), Geert De Neve (University of Sussex), Jumpei Ichinosawa (Miyagi Gakuin Women's University), Carolyn Lesorogol (Washington University), Ty Matejowsky (University of Central Florida), Atsuro Morita (Osaka University), Lionel Obadia (Université Lyon 2), Noel B. Salazar (University of Leuven), Cynthia Werner (Texas A&M University), and Tamar Diana Wilson (University of Missouri). Finally, one more development: this is the first REA volume to focus exclusively on the phenomenon of climate and environmental change, one from which – it seems safe to say – no human living today is completely unaffected.

The aim of this volume is not to critically examine the politics of climate change – to lay blame, to analyze denials, or to expose cover-ups. This has already been eloquently and convincingly done by a variety of authors (Hoggan, 2009; Mann, 2012; Oreskes & Conway, 2010). After all, despite the stubborn persistence of conflicting views, human culpability for global warming, the biggest attention-getter element of the general phenomenon of climate change, has already been proven beyond all reasonable doubt: "It is all but certain that human activity has caused a steady increase in global temperatures over the past 60 years, leading to warmer oceans and an acceleration in rising sea levels" declared a recent U.N. report (Washington Post, 2013). Adding to our worries over what we've collectively done to our world and ourselves, archeological research points to an increasing likelihood that climate change was a major factor in the demise of numerous ancient human civilizations (APF-JIJI, 2013a; Fiske et al., 2014).

The aim of the present volume, rather, is to contribute to a more thorough understanding of the ways in which people around the world have adapted (or struggled or failed to adapt) culturally to changing economic conditions caused by climate change and/or shifts in environmental quality. It does this by focusing primarily on specific situations in particular locations, showcasing and confirming the strength and value of intensive ethnographic investigation.

Part I of the volume is comprised of three papers focusing on situations in three separate parts of the world: Nepal, Reunion Island (in the Indian Ocean), and southeastern British Columbia, Canada. First here, Pasang Yangjee Sherpa analyzes institutional efforts at adapting to the effects of climate change in her own native land of Nepal – a land which, in addition to coping with these issues, is struggling to recover from a devastating earthquake at the time of this writing. One main point that arises from her essay is the finding that "climate change" may mean something different to different people living in the same general area, even if said people constitute a single "people." By focusing on four different activities, Sherpa demonstrates a need for a deeper understanding of local situations and culture – the local cultural world, no less – and for more sincere and effective outreach efforts on the part of institutional actors, in order to make meaningful contributions to climate change-related problems.

Next, Sophie Bouly de Lesdain raises a similar point by shining a light on decision-making processes among small-scale photovoltaic energy producers on Reunion Island, a distant French overseas department in the Indian Ocean, some 675 km off the eastern coast of Madagascar. On the one hand, there is a desire for greater independence from the local power grid. After all, electricity-generating fuels must be imported. This seems to have given rise to a fair degree of concern about household consumption levels. But there is also a more immediate, practical reason to put solar panels on one's roof: frequent power outages that can last for some days. In addition, government policy – in the form of tax laws, direct support, and artificial electricity price adjustment – emerges as a major factor. The last of these has resulted in electricity rates that are equal across the board in all parts of France and its overseas departments. Uncertainty over the future of these policies, it is shown here, has contributed greatly to residents' choices to aim for energy independence. Importantly, pointing to the value of ethnological research on these kinds of global/state/local processes, the author reveals decision-making that is unexplainable by simple cost/benefit analyses. Concerning the recent proliferation of solar power initiatives around the globe, Sophie Bouly de Lesdain's findings here will no doubt prove to be applicable to a variety of similar situations.

Finally, this section of the volume closes with an essay by Patrick B. Patterson – who also contributed a paper to volume 26 of REA – on the effects that climate change are having on local perceptions of forest resources and work in British Columbia, Canada. Again, uncertainty comes to the forefront, as the future of a formerly highly dependable resource and centerpiece of the regional economy – the forest – becomes murky due to increasing pine beetle damage caused by a general warming trend. Worry over ability to continue serving as stewards of the forests, and even over the future of the industry itself, point to a host of challenges that climatic shifts appear to hold in store for human societies all over the world.

Part II of this volume continues Part I's focus on the local, but the five papers that comprise it are tied together by a common concern with food production and/or procurement – a vital aspect of human adaptation to climate change, as warnings of food crises caused by the phenomenon have been sounding for quite some time now (e.g., The Observer, 2013). The first, by Ana Maria de Souza Mello Bicalho and Scott William Hoefle, does not focus exclusively on food production – its concern with logging provides continuity with the closing paper of the previous section of the volume – but frontier peasant smallholders in and near conservation units in the Central Amazon constitute an important factor in the overarching question here of "whether new full-time or multifunctional part-time rural activities can significantly contribute to Amazonian livelihoods." As is generally known, in addition to our planet's polar regions, the Amazon basin has long served as a powerful symbol of all that is ecologically wrong with human society today. On a concrete level, rapid deforestation coupled with rapid urban growth there presents especially difficult problems for Brazil's government and those individual and institutional actors who would seek to preserve this sensitive environment (Romero, 2012). Here, "multifunctional" refers both to local adaptive strategies combining agricultural production and nonagricultural activities, and also to the research strategy employed by Bicalho and Hoefle. Ultimately, the goal is to help find a way to balance conservation with antipoverty measures in order to reverse deforestation and reduce (often violent) land disputes – a lofty one indeed, but one that might be achievable through successful multifunctional approaches.

The second and third papers of this part of the volume both remind us of the importance of hands-on, empirical research into the intersections of atmospheric events and local experience and knowledge if one is to fully comprehend the effects of climate change on human societies. They also both highlight specific crop choices made by local producers and give some insight into the temporal vicissitudes of producers' underlying decision-making processes. First, Olivia Barnett-Naghshineh reminds us that conceptions of climate change – the discourse itself being a Western import in her geographical research locus of Papua New Guinea – are always subject to local cultural ontological reworkings. In other words, climate change discourses must take on local meanings for them to become locally relevant. But at the same time, in this case, local climate change ideas do not mesh in all respects with actual changes in crop production and exchange, as the food landscape concerned here evolves, along with the development of the market. The author's point that local people are not mere victims of climatic shifts but rather agency-possessing actors in the unfolding drama deserves mention. Next, Jane I. Guyer revisits a long-term research site in Nigeria through previously collected ethnographic data to determine the ways in which shifts in the region's climate have impacted farmers' decisions on crop type and planting schedules, also taking into account market and technology changes. She finds that uncertainly regarding the weather has had a considerable impact. Just as Barnett-Naghshineh identified impacts of climatic and market transformations on food exchanges in Papua New Guinea, Guyer sees impacts of these on both subsistence strategies and monetary relationships in western Nigeria, indicating a possibility of increased social tensions caused by changes in climate.

The fourth and fifth papers of Part II are both heavily concerned with water – the first with shortages of it (drought) in southern China and the second with protein drawn from it off the coast of the island of Flores in Indonesia. In the China case, Meine Pieter van Dijk and Hao Li find that although there is a tendency for state policies to fail to have meaningful impacts "on the ground" so to speak, local producers – both individually and collectively – employ a variety of adaptive strategies to deal with drought conditions to sufficiently irrigate their fields and provide their crops with water. The researchers identify problems such as a propensity for local elites to retain control, and also a common aversion to risk, but also find hope for better policy formulation and implementation at the local (rather than at the national) government level. They call for greater decentralization in China to deal better with changing climate conditions. In the final paper of Part II, Victoria C. Ramenzoni finds that small-scale fishermen are now less able than before to successfully practice their craft while relying on traditional lunar calendars to determine when and where to do so. This is important because it threatens to upset the balance between natural (common-pool) resources and societal needs for cash income and protein from the sea. Major implications of Ramenzoni's article include: (1) intense ethnographic research can elucidate complex micro-processes that wide-ranging government surveys miss, and (2) increased inability to effectively deal with mounting uncertainty may lead to over-exploitation of critical marine resources if adequate preventative steps are not taken.

Part III consists of three papers that are linked by a shared concern with climate change effects in urban environments. Monica L. Smith, also a contributor to volume 32 of REA, kicks off the last section of this volume with a paper that reaches back several thousand years to tackle the very broad issue of the emergence of the concept of sustainability in human societies – quite possibly to deal with diminished options for dealing effectively with climatic shifts due to sedentary life with increased population densities. But in these environments, according to Smith, the need to think about long-term sustainability became obscured by provision networks that reached farther and farther from the cities to bring needed items to them. Furthermore, Smith suggests, the human mind is wired now more for thinking about and dealing with short-term risks than for long-term ones (largely due to our social evolution) and that, therefore, the problem of sustainability for the (distant) future – although important now for obvious reasons – is not one that we are naturally equipped to tackle. Smith's insights may shed some light on interminable attempts to discredit climate change theories and findings.

In the second paper here, Ty Matejowsky – who has contributed numerous papers to REA in the past and who also co-edited volume 32 – explores retailers' coping strategies in the face of recurrent, and increasingly large and destructive, typhoons in the Philippines. This is a problem that can only be expected to worsen in the coming decades, as experts increasingly point to global warming as a culprit behind extreme weather events (APF-JIJI, 2013b). Specifically here, Matejowsky concentrates on small- and medium-scale enterprises (SMEs) – which comprise the overwhelming majority of business enterprises in the country – in the northern city of Dagupan, Pangasinan province. Considering the fact that the Philippine Islands are especially prone to natural disasters of various kinds, one might expect SME proprietors of the country to rely heavily on insurance policies, for example, but Matejowsky finds surprisingly little of this. Neither does appealing to the state for financial assistance emerge as a major strategy. Rather, personal networks and simply moving merchandise upward (i.e., farther from the water line) inside stores take precedence as major coping mechanisms within a general atmosphere of fatalistic ambivalence. Human – including supplier – networks seem to be deserving of greater attention in research on increased uncertainty in the wake of climate change-related adverse events in the Philippines.

This volume ends with a look by Richard K. Reed into the effects of human-caused environmental destruction on indigenous peoples in Paraguay, and particularly as these can be seen in the country's capital city of Asunción. As Reed explains, rapid and massive deforestation over the last few decades has displaced many native populations inside the country, and the majority of the people have sought refuge in the cities. Reed focuses on the Mbyá-Guarani in Asunción, and finds that they have not adapted well to city life; instead, as a people, they appear to be engaging in a form of resistance to the state and to the urban environment. Case studies such as this should grow in importance as more and more climate change refugees appear in the coming years and decades.

In closing, it seems prudent to call attention to several major points that have been raised by the papers in this volume, as dealing with – and any hope of reversing – climate change at the local level will depend on comprehending them. First, there is the issue of coping with uncertainty created by changes in local climates – a problem that tends to become intensified by the influx of information about similar developments in other parts of the world, and by the global issue of rising sea levels. The papers here have demonstrated a variety of ways in which people in different parts of the world are dealing with uncertainty as they try to cope with climate change and environmental degradation. Next, the value of local knowledge in facing change and in finding ways to prevent it has been highlighted. Third, research here has demonstrated that the climate change discourse itself – often a Western import, or perceived as one – must be adapted to local situations in order to have relevancy. Finally, some papers here have alluded to possible solutions to ecologically detrimental practices that have been identified as causes of climate change. With both coping and rectifying in mind, the American Anthropological Association launched the Global Climate Change Task Force in 2010, and published this group's final report in December 2014. The findings of this volume of REA mesh well with and complement those of the AAA document, which is freely available for download on the AAA website. Interested readers are encouraged to consult it for a greater insight into anthropological perspectives on climate change.

Donald C. Wood
Editor

REFERENCES

APF-JIJI. (2013a, August 16). Climate change linked to ancient civilizations' fall. *Japan Times*, p. 10.

APF-JIJI. (2013b, September 7). Global warming tied to wild weather. *Japan Times*, p. 7.

Fiske, S. J., Crate, S. A., Crumley, C. L., Galvin, K., Lazrus, H., Lucero, L., … Wilk, R. (2014, December). *Changing the atmosphere: Anthropology and climate change*. Final report of the AAA Global Climate Change Task Force (137 pp.). Arlington, VA: American Anthropological Association.

Hoggan, J. (2009). *Climate cover-up: The crusade to deny global warming*. Vancouver, BC: Greystone Books.

Mann, M. E. (2012). *The hockey stick and the climate wars: Dispatches from the front lines*. New York, NY: Columbia University Press.

Oreskes, N., & Conway, E. M. (2010). *Merchants of doubt: How a handful of scientists obscured the truth on issues from tobacco smoke to global warming*. New York, NY: Bloomsbury Press.

Romero, S. (2012, November 25). Swallowing rain forest, cities surge in Amazon. *New York Times*, p. 5.

The Observer. (2013, April 20). Climate change feared to create global food crisis. *Japan Times*, p. 7.

Washington Post. (2013, August 22). Humans '95% likely' behind warming. *Japan Times*, p. 4.

PART I
LOCAL CULTURAL ADAPTATIONS/ ADJUSTMENTS

INSTITUTIONAL CLIMATE CHANGE ADAPTATION EFFORTS AMONG THE SHERPAS OF THE MOUNT EVEREST REGION, NEPAL

Pasang Yangjee Sherpa

ABSTRACT

Purpose – *The purpose of this paper is to show the complexity in dealing with climate change adaptation at the local level, and to show how social and institutional factors in addition to the ecological challenges contribute to that complexity.*

Methodology/approach – *This paper examines four institutional climate change activities and reveals how institutions currently address climate change, and how the Sherpas are involved in the process. It draws on three sorts of material: the interviews and observations conducted during my field research in 2010 and 2011; my personal experiences as a Sherpa woman; my recent participation in Sherpa face-to-face and online communities.*

Findings – *Organizing institutional climate change activities to draw international attention alone are not sufficient to address climate change*

Climate Change, Culture, and Economics: Anthropological Investigations
Research in Economic Anthropology, Volume 35, 3–23

ISSN: 0190-1281/doi:10.1108/S0190-128120150000035001

adaptation issues. Communities at the local level cannot be assumed to be homogeneous entities. Institutional climate change adaptation efforts cannot assume that by reaching out to a few individuals in the region they will benefit the whole. Institutional activities have increased receptivity to scientific climate change knowledge, but it has also increased fear of an impending doom, and anger over the continuous discussion of climate change without concrete actions.

Research implications – *Future research in the Everest region should include residents from all ethnic groups considering their historical contacts and interactions.*

Originality/value – *It is crucial that not only the Sherpa agency (or lack of agency) or understandings are examined but the institutional engagements and delivery are also assessed to practically, effectively, and sustainably address the challenges of climate change adaptation.*

Keywords: Nepal; climate change; culture world; social heterogeneity; Sherpas; Mount Everest region

In the Everest region, climate change is not only an environmental issue it is also institutional. Climate change is understood locally as an issue associated with institutions. Institutions, discussed in this paper, are formal groups and organizations with official way of working, small and large, and based locally and non-locally. Institutions engaged in climate change activities in the Everest region are NGOs, government offices, and commercial entities. Institutions introduced and apply the phrase climate change in their activities designed as responses to local effects of a global phenomenon (Sherpa, 2014). This paper examines four such climate change activities and reveals how institutions currently address climate change, and how the Sherpas are involved in the process. The goal of this paper is to show the complexity in dealing with climate change adaptation at the local level, and to show how various social and institutional factors in addition to the ecological challenges contribute to that complexity.

Global scale analyses of adaptation alone "cannot capture the complexity and diversity of changes that are already taking place in response to climate change, nor can they capture the significance of losses that are already being experienced" (Adger, Lorenzoni, & O' Brien, 2009, p. 2). Local level analyses of adaptation are complementary to the global scale analyses in understanding existing resources to respond to climate change, and the magnitude of losses already being experienced. Local level analyses of adaptation are also useful in understanding the multiple socio-economic stressors that further assist or deter climate change adaptation efforts.

"The limits to adaptation, as a response to climate change, depend on ecological thresholds, individual and cultural values, and institutions and governance. As these social, physical, and ecological factors together will determine whether adaptation is successful, analyses that overlook one aspect may present a dangerously misleading understanding of the consequences of climate change" (Adger et al., 2009, p. 2). Therefore, I argue that to understand climate change impacts on people at the local level, from an adaptation perspective, it is crucial to understand their cultural world.

"Cultural worlds are the most extensive social environments where people routinely interact to secure their existence, and where in spite of linguistic or ethnic differences, people share understandings about the nature of the world, and about common human rights and obligations" (Bodley, 2003, p. 7). Cultural worlds are not fixed spatially, but space can be a source of identity that brings people together. They are socially dynamic, involving people of different backgrounds, and sharing understandings, spoken or unspoken, about the nature of the world, and about common human rights and obligations.

I further argue that in the small-scale Sherpa cultural world based in the Everest region, institutions are one of the many stakeholders, and the residents themselves are socially heterogeneous, not the homogeneous communities which institutions treat them as. I argue that considering Sherpa social heterogeneity is crucial to understanding the complex Sherpa agency upon which the effectiveness of any institutional climate change adaptation efforts depend. It is necessary to explore social heterogeneity and multiple interests, rather than equating community with place (Waylen, Fischer, McGowan, & Milner-Gulland, 2013, p. 578). Heterogeneity is a characteristic of small-scale cultures and a corollary to subsidiarity (Bodley, 2012, p. 13). Subsidiarity means that decisions are made at the lowest feasible level in the social structure, starting with the individual, then moving up to the household, family, village, and regional levels as demanded by the circumstances. It encourages people to mobilize local social networks to solve problems, rather than relying on foreign, higher level hierarchies (Bodley, 2012, p. 13). Social networks are neither static nor uniform. During my fieldwork, I found that social networks for the Sherpas included neighbors, who may or may not be relatives. Similar socioeconomic status was found to be a significant criterion for inclusion in these networks. Close relatives with uneven socioeconomic statuses, although living nearby, were not found to interact in the same social networks (Sherpa, 2014).

The Sherpa cultural world includes Sherpas and other ethnic groups. In this paper, I take the Sherpa cultural world as my vantage point for two reasons: (1) my ethnographic research and engagement have focused on the Sherpa people; and importantly for this paper, (2) institutions target their climate change activities on the Sherpas in the Everest region. I draw on three sorts of material: the interviews and observations conducted during my field research in 2010 and 2011; my personal experiences as a Sherpa woman; my recent participation in Sherpa face-to-face and online communities. The field research that informs this paper is based on 15 months of study of Sherpa perceptions of climate change in the Mount Everest region of Nepal (Sherpa, 2014). Here, I begin by discussing the Sherpa cultural world and Sherpa social heterogeneity. Then, I explore national climate change adaptation mechanisms along with institutional engagements at the local level to better understand how and why institutions are part of the Sherpa cultural world. This is followed by an examination of four institutional climate change activities that show the complexity in dealing with climate change adaptation at the local level by revealing the presence of multiple stakeholders, their multiple interests, and their varying levels of agency.

SHERPA CULTURAL WORLD

We call ourselves "Sharwa" (the Easterners) in the Sherpa language from which the word "Sherpa" has derived. "Eastern" here means eastern Himalayan region. "Sherpa" is now the name of our ethnic group: the way we introduce ourselves; and the way we are recognized. We have our own distinct language, religious belief systems, and cultural practices (Fisher, 1990; Ortner, 1989; Von Furer-Haimendorf, 1964). We practice Nyingma Buddhism and believe that the Everest region is a Beyul, a sacred hidden valley blessed by Padmasambhava that can provide refuge in times of distress (Sherpa, 2013; Skog, 2010; Spoon & Sherpa, 2008).

Sherpas are popularly associated with the Everest region and the Himalayas in Nepal. Sherpas today, however, live in different parts of the world. Participant observation of community events suggests that after

Kathmandu, the capital city in Nepal, the second largest concentration of the Sherpas is in Queens, New York area. Online social sites such as SherpaKyidug.org and Facebook have allowed Sherpas to stay connected with each other. The Sherpa Kyidug website was created and is managed by the Sherpa community in Queens. At these online sites, we share pictures of our family and individual accomplishments. We also comment on issues and topics ranging from loss of language, religious rituals to financial aids for students, and real estate markets. Regardless of where we live, our villages in Nepal continue to be a major source of our identity.

Locally, the Everest region is known as Khumbu, the northern part, and Pharak, the southern part. This region is also the territory of the Sagarmatha (Mount Everest) National Park (SNP) and Sagarmatha National Park Buffer Zone (SNPBZ). Khumbu lies within the territory of SNP and Pharak (the southern part of the region) lies within the territory of SNPBZ. Villages in Pharak are lower in elevation and warmer in temperature than Khumbu villages. This allows diverse flora and fauna to thrive in Pharak compared to the alpine conditions of Khumbu. According to the 2011 census, there are 968 households and a population of 3,709 in Pharak. In Khumbu, there are 1,031 households and a population of 3,452 (Government of Nepal, 2014).

Traditionally, the Sherpa economy involved agro-pastoralism and trans Himalayan trading. After the successful Everest ascent by Tenzing Norgay Sherpa and Sir Edmund Hillary, further facilitated by an operating airport in the Pharak village of Lukla, the region witnessed an influx of tourists. The tourism industry that began in the 1950s with a few Westerners now supports more than 35,000 tourists annually. The local economy today depends largely on the tourism industry. In 1991, Barbara Brower found that every household in the Everest region was either directly or indirectly involved in the tourism industry (Brower, 1991). Sherpas were directly involved as porters, guides or shopkeepers and teahouse owners, and indirectly as vegetable sellers and laborers due to the restructuring of the local economy. As tourism flourished, individuals and families from neighboring regions migrated into the Everest region in search of better economic opportunities. The local population in the region is still predominantly Sherpa but there is an increasing population of other ethnic groups.

On-route villages on the main trail to Mount Everest have businesses catering to tourists, facilities to support the tourism industry and modern amenities to attract visitors. These villages are visibly different from the off-route villages, which look more traditional. Households in off-route villages depend on traditional agro-pastoralism with new markets for their crops in on-route villages. A couple of successful greenhouses in off-route villages have also started providing vegetables and tomatoes to on-route lodges.

Walking distances for adjacent on-route and off-route villages are normally half an hour to 45 minutes. Off-route villages receive fewer visitors and tourists compared to the tens of thousands that travel through the on-route villages. Institutions also organize their activities in on-route villages. On-route and off-route villagers therefore have different opportunities, in terms of kind and number, for socioeconomic interaction with each other and with visitors.

During my field research, I found that while tourism is still the most dominant part of the local economy, transnationalism has also become significantly prominent. Every family in Pharak had at least one close relative or a neighbor in the United States. One of the villages in Pharak was found to have at least one member from every household working in the United States. The sponsors for the annual Dumje festival celebration in Pharak, from 2006 until 2014, were families with at least one person either currently living in or had visited the United States in the past. The cost of sponsoring the Dumje festival can run anywhere from US $3,000 to $5,000, an expensive undertaking for the local residents, and virtually impossible for someone who does not have transnational support.

Sherpas today are employed in diverse occupations. Schools built by Sir Edmund Hillary's the Himalayan Trust in the 1960s paved the way for Sherpas from the Everest region to receive education. Families that migrated to Kathmandu in the 1970s and 1980s and those that could afford to, after tourism flourished, sent their children to English-medium schools in the city. Several Sherpas are now attending or have graduated from Western universities, and are working in different business and service sectors.

SOCIAL HETEROGENEITY AND CLIMATE CHANGE

In Pharak, I found that Sherpas perceived climate change differently (Sherpa, 2014) depending on the socioeconomic backgrounds of individual Sherpas. Sherpas were exposed to different environmental changes and understood these changes through different knowledge systems based on their gender, age group, occupation, and residence in on-route or off-route villages. Education levels did not appear to affect their perceptions of climate change as it did in Khumbu, which Spoon found in his study of traditional ecological knowledge (Spoon, 2011). They were also likely to change their perceptions based on what other members of the society, including institutions and the media said.

Environmental changes described were changes in rainfall, snowfall, temperature, arrival of new insects, and crop diseases. Informants also talked about the extreme events such as glacial lake outburst floods (GLOFs), heavy rain flooding, strong winds, and earthquakes. Informants described these changes from a scientific perspective usually following institutional narratives and also from a religious perspective following local belief system of the need to appease deities so the land and the people could be protected. These different ways of understanding environmental changes were not seen as conflicting. One of my informants trained as a Buddhist monk and also well versed in institutional narratives of climate change explained that while science is logical, religion is necessary.

In my research, I observed that a female farmer in her 30s from an off-route village was aware of new crop diseases showing up in her village, changes in the flowering time for different plants, and arrival of birds at different times. A male herder in his 60s had not heard about the phrase "climate change" before and was not aware of various climate change activities that have been organized in the region. He was, however, aware of the national park policies that restricted him from moving his cattle from one community forestry user group to another. He was also aware of the movements of wild animals in the forests and the changes in the temperature that affected his seasonal movements to higher and lower elevations. A female in her 20s, who works as a household helper collecting firewood described how the forests are changing and how she has to walk to different places to collect firewood and grass for cattle. She was also able to describe different kinds of edible wild plants, which her peers working in the lodges were not able to do.

Participant observation of the institutional activities and further discussions with key informants revealed that a group of male hotel owners from on-route villages in their 30s–40s regularly attend institutional climate change activities. It was found that this group was fairly inundated with information about climate change as understood and disseminated by the institutions organizing climate change activities. They were however less familiar with the environmental changes than the female farmer in her 30s, the male herder in his 60s, or the female helper in her 20s.

The residents who did not participate in the institutional activities, which constituted a majority of the population, were unaware of the institutional climate change narratives. They were, however, more knowledgeable about the environmental changes. Their participation in the institutional activities was limited because attending them meant taking hours and sometimes days out of their busy lives since most of the activities were conducted in on-route Khumbu villages.

CLIMATE CHANGE ADAPTATION IN NEPAL

The government of Nepal recognizes the rising trend in temperature and an erratic rainfall pattern that are scientifically recorded for the country and associated with natural disasters such as landslides, floods and drought, including glacial lake outburst floods (GLOFs) (Ministry of Environment, 2011). The resulting damage and destructions caused by such climate change induced natural disasters make the climate change risks for Nepal. In order to deal with climate change risks, the government has put in place the National Adaptation Program of Action (NAPA) and Local Adaptation Plan of Action (LAPA). There is also a mandatory provision for disbursing at least 80 percent of the available climate change budget for the implementation of adaptation and climate change activities at the local level (Ministry of Environment, 2010).

The LAPA was primarily initiated to identify local level climate change risks and adaptation needs with the participation of people at the local levels throughout the country, as prescribed by the NAPA. While the LAPA concept was conceived to bring forth the climate change vulnerabilities and action needs of the people at the local level, the tools for LAPA development and implementation were crafted at the national level. In 2011, the National Framework for LAPA was formulated to assist in the development and implementation of action plans, including support for the integration of climate change adaptation and resilience aspects in local and national plans through identified entry points – agriculture, forestry, health, water and sanitation, watersheds and microfinance.

The LAPA documents serve as guidelines to allocate and direct resources for future climate change adaptation activities. Therefore, they are significant instruments for climate change adaptation in Nepal. Several regions throughout Nepal already have a prepared LAPA document. The document for the Everest region titled "Khumbu LAPA" was prepared in December of 2013 by the High Mountain Adaptation Partnership (HiMAP) with support from the United States Agency for International

Development (USAID) (High Mountains Adaptation Partnership [HiMAP], 2014). As of June 2014, the Khumbu LAPA document was, however, yet to receive support from the SNPBZ Management Committee, formed with local Sherpa representatives, due to their lack of inclusion in the process of developing this document.

INSTITUTIONAL CLIMATE CHANGE ACTIVITIES

In conversations, institutions are referred to as "office work" or "project work." They are also referred to with acronyms of individual institutions such as WWF (World Wildlife Fund), SPCC (Sagarmatha Pollution Control Committee), and TRPAP (Tourism for Rural Poverty Alleviation Program). Since the 1960s, institutions have pursued development, conservation efforts, and now climate change in order to assist the Sherpas. Institutional climate change activities in the Everest region began in the new millennium with WWF's "Climate Change Witness" project in 2004. This project provided an international platform for three Sherpas, two men and a woman, to tell their stories of how previous GLOF events disrupted their lives.

Institutional climate change adaptation efforts as they pertain to the Everest region originate largely but not exclusively from two sources: (1) governmental and (2) nongovernmental organizations. Scientific studies of the physical environment in the Everest region have been ongoing for several decades (Byers, 2007; Shrestha & Aryal, 2011; Somos-Valenzuela, McKinney, Byers, & Rounce, 2013; Watanabe, Kameyama, & Sato, 1995). These studies inform current institutional adaptation efforts.

Fig. 1 shows the distribution ($n = 28$) of institutional climate change programs and activities conducted between 2004 and 2011 in the Everest region and/or involving the Sherpas. Virtually all of these institutional activities were organized to "raise awareness" of the Sherpas about the potential Imja GLOF and the global phenomenon of climate change as described by the Intergovernmental Panel on Climate Change (IPCC), and to "build capacity" of the local Sherpas in dealing with the potential Imja GLOF by equipping them with scientific knowledge and communication technology. These activities were also organized to draw international attention to the Himalayan melting of glaciers and the threats posed to the Sherpas living in the vicinity.

As shown in Fig. 1, between 2008 and 2009, there was a surge of institutional activities in the Everest region, which declined sharply after the

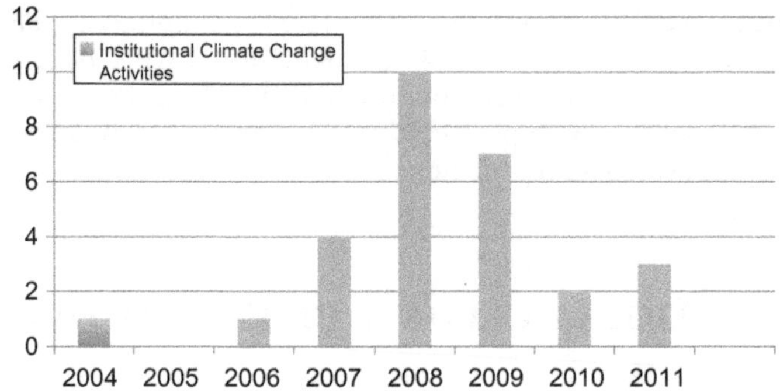

Fig. 1. Institutional Climate Change Activities.

event. A survey of the activities organized in these years show that while the Sherpas and the Everest region were targeted for climate change awareness raising adaptation efforts, the activities themselves were designed to produce knowledge of the dire situation of a popular mountain people living in a popular tourist destination to present at an important international climate summit, the COP 15. Rather than capturing the wide range of climate change effects being observed and experienced by the Sherpas, these activities served as opportunities for the institutional organizers to showcase a climate change narrative of the Himalayas – focused on glacial melting and GLOFs – at an international stage. Today, a wide range of climate change effects and their impacts on the people still continue to exist unexamined and possibly magnified. Further examination of these institutional activities also reveals how institutions currently approach climate change issues indicating what the future for climate change action may look like, if the current trend continues.

Below, I examine four different cases of institutional activities organized in 2008, 2009, 2010, and 2011 for the Sherpas in the Everest region. These cases show the kinds of climate change activities that are organized in the region and reveal how the Sherpas are participating. I chose these activities as my case studies here because these were the ones my informants mentioned regularly during my fieldwork and these were also found to be the ones that had the most impact on my informants. Although all of these activities were initiated with positive intentions, as discussions below show sometimes there were unintended negative consequences too. Each case begins by describing the activity followed by how Sherpas responded to them. The responses show the complex agency of the Sherpa residents, which is often overlooked while institutional activities are designed, developed, and implemented.

Case I: Adaptation to Climate Change and Increasing Resilience of Local People in Khumbu

In 2008, the International Center for Integrated Mountain Development (ICIMOD) in collaboration with the SNPBZ Management Committee organized a workshop entitled *Adaptation to Climate Change and Increasing Resilience of Local People in Khumbu.* Participants in this workshop represented locally based institutions – community forest user groups, youth clubs, eco clubs, governmental administrative office, national park office, SNPBZ management committee – and Kathmandu based institutions, such as the Himalayan Trust, ICIMOD, and Gramin Pahunch (*Rural Reach*). Around half of the participants were full time residents and the other half was composed of the organizers, non-resident Sherpas, and temporary government staff.

This awareness workshop was organized to strengthen local communities' and organizations' ability to adapt to climate change. It was held in Namche Bazar, a popular Khumbu village. At this workshop findings from scientific studies that examined climate change impact on glaciers and glacial lakes in Dudh Koshi (the major river in the Everest region) basin were shared. A demonstration of real-time monitoring of Imja glacial lake using geo-ICT tools, established in cooperation with the Department of National Parks and Wildlife Conservation (DNPWC) Nepal, Keio University Japan and ICIMOD, was conducted. This information sharing workshop and the interaction with the local people and the organizations were believed to help in developing a possible early warning system that would reduce risks of potential GLOFs.

In 2010, during my preliminary research in Pharak, informants said that they were concerned about the potential Imja GLOF. Pema, a key informant, was one of them. Her village is at a lower elevation and closer to the Dudh Koshi in comparison to other neighboring villages. She was young when Dig Tsho flooded in 1985 and does not know much about it but remembers how frightened people were. After learning about the possibility of the Imja GLOF, she was further encouraged to send her daughters to Kathmandu to attend schools there, fearing for their safety. She also asked her mother to stay in Kathmandu and not to return to the village.

She explained, "My husband and I have to be here in the village. We cannot live in Kathmandu with our children because this is where we work and earn money. Here, I have a lodge and tourists come. In Kathmandu, I cannot do anything. I am not educated. Maybe if I were in the United States [like my relatives] I could make good money, but it is difficult to go there too."

During interviews, informants described an incident when they received a false alarm about the Imja GLOF. They had to flee from their homes in search of safer grounds. In early 2011, Rinzin, another informant described, "The villagers in Thumbuk [which is close to the Dudh Koshi] had received some news of rising water level in a Khumbu stream. This was then interpreted to be the Imja GLOF. There had been rumors about the potential GLOF in Pharak spread by the participants of the workshop in Namche Bazaar, a week before this incident." He added, "I was one of the participants at this workshop and I knew that Imja was not coming. So, I was not worried but I joined my fellow villagers anyway … I thought it was funny when my relative called me to tell me I should find safer ground for myself and my family when he himself was on the bridge. I told him if there is going to be a real GLOF, he should not be on the bridge. Then, I asked him if the river below him had risen and he replied, no."

Daki had given birth to a child a couple days before the false alarm. When her in-laws heard about the GLOF, all of them fled to a higher place behind her home with the newborn. It was a cold, rainy day and her refuge was the potato field. She stayed there the whole night with the child. Pema later said, "Fortunately, nothing happened to her or the baby. They seemed fine."

Pema, herself, was attending a cousin's wedding in a different village. She recalled the next morning when she returned home to find everyone shaken by the incident. Her neighbors and she laughed about the false alarm after everyone realized that nothing had actually happened. "Good thing I wasn't here, I would have not known what to do," sighed Pema.

While ICIMOD believed that the information sharing and the interaction program that took place in Namche Bazaar would help the Sherpas develop warning systems and reduce the risks of potential Imja GLOF, as they discuss it in their reports, what was not intended was the spread of rumors potentially putting people's lives in danger.

Sherpas in Pharak, who would be hit hard if the Imja GLOF occurred, were unaware and misinformed about the situation. In my follow up interviews three years after the workshop, I did not find Sherpas in Pharak to be more informed about scientific findings of climate change despite several institutional climate change activities had already been carried out in the region.

Case II: Nepali Cabinet Meeting in Kalapatthar

In December of 2009, the government of Nepal held a cabinet meeting in Kalapatthar (17,000 ft.) next to the Everest base camp. Twenty-one ministers traveled by helicopter to attend this meeting. The objective of this meeting was to draw international attention to the rapid melting of the Himalayan glaciers due to global warming. Mr. Madhav Kumar Nepal, the Prime Minister at the time, stressed that the Himalayas are "a global treasure," which needed to be protected (BBC, 2009). During this meeting, the ministers also endorsed the Sagarmatha (Everest) declaration on climate change, where the government of Nepal expressed its commitment to work together with other nations to protect the Himalayas. Several national and international media covered this event. An article on the BBC website mentioned, "Crowds of local people, many of them wearing traditional outfits, gathered at the village of Syangboche, from where the ministerial helicopters took off." The trip to Kalapatthar was funded by a group of Nepali private businesses most of them in the tourism sector.

Before the cabinet meeting in Kalapatthar, Sherpas from around the world discussed the relevance and significance of this meeting to the Sherpas, especially of the Everest region. This discussion occurred on the Sherpa Kyidug website. Sherpas engaged in the discussion agreed that the meeting should recognize the indigenous Sherpa people and their customs, if the meeting were to be held in their homeland. A list of issues (see Table 1) raised in this discussion forum ranged from climbing permits for mountains in Solukhumbu district to avoiding killing livestock for meals for the ministers (Sherpa Kyidug, 2012).

After the cabinet meeting concluded, an individual identified as *Khumjung*, who had actively participated in the discussion, wrote on the Sherpa Kyidug discussion thread:

> If the whole show was to attract world's attention, then it was a complete failure. Unlike the underwater cabinet meeting in Maldives, which was on the front page, and TV of every news outlet, the cabinet meeting in Nepal was nowhere to be seen … I started with New York Times … [Then] I typed "Nepal" and searched [online] … that brought couple of articles … nothing on the "Cabinet Meeting." Then I checked the Wall Street Journal … nothing on the front page, nothing on "World" Section, nothing on "Asia" Section, and nothing when you search "Nepal," although there was a nice

Table 1. List of Issues Raised in the Sherpa Kyidug Discussion Forum.

Issues
1. Revenue collected from issuing climbing permits for mountains in Solukhumbu district should be spent on local development projects such as education, health care, and infrastructure development.
2. Local Sherpas should be invited for participation and consultation in all research and study, especially of the potential Imja GLOF.
3. Sherpa leaders should also be invited to endorse the declaration on climate change.
4. Tengboche Rinpoche, the spiritual leader of the Everest region, should be invited as the Chief Guest for any associated events. The people of Solukhumbu should be allowed to host the event.
5. Killing of livestock for meals to serve during the event should be strictly prohibited to maintain the sanctity of the Beyul Khumbu.
6. The fame of Sherpas should not be treated as a trademark to collect international funds.
7. The government of Nepal should restrict construction of hotels and lodges in the national park areas, especially by outsiders.

article about melting glaciers in Nepal. Upset, but undaunted, I went to CNN. Again, nothing on the front page, and after selecting the "WORLD" tab, nothing on Nepal jumped out. But … in small fonts, at the bottom right of the page, it said "Everest hosts Nepal cabinet meeting." Well, it wasn't on the front page, and unless you specifically and diligently look for the article, you will probably not find it.

When asked about the cabinet meeting, which was organized in Khumbu, most of my informants from Pharak including Temba were not sure what had happened during the meeting and some of them even said they learned about the meeting after watching the news on television, which broadcasts from Kathmandu. An educated woman in her 30s from Lukla was aware of the arrival of ministers in the region but she mentioned that the interaction of the local people with the ministers was minimal. A male hotel owner from Namche in his 50s noted that since the meeting was conducted to draw international attention and there was media coverage, it was successful. However, the hotel owner added that the cabinet meeting did not include local issues and his villagers expressed their discontent to the ministers at the Syangboche airport. An educated male from Khumbu in his 20s thought the cabinet meeting was not coordinated properly with the local people and it was not effective in delivering the critical messages of climate change in his village.

Case III: The Imja Tsho Action Event

The 2009 Imja Tsho Action Event was organized by ICIMOD and Initiatives for Development & Eco Action Support (iDEAS), a nonprofit organization based in Kathmandu, in collaboration with the Sherwi Yondhen Tshokpa (SYT), a local Sherpa students' group. This event included an inter-school art competition, a letter writing competition, and a footrace. According to ICIMOD (2012), "the objective [of this event] was to encourage children from the Khumbu region to express their views on the possible impacts and effects of climate change; to raise awareness of climate change effects among the mountain community of the Khumbu region; and to bring the 'voices' of mountain children from Khumbu to the COP 15 in Copenhagen, 7–18 December 2009." The entries for the inter-school art competition and the letter writing competition were used as climate change information sharing materials on ICIMOD's website and the COP 15.

According to Dawa Steven Sherpa, climate change ambassador and the founder of iDEAS, the "Beat the GLOF Action Run" was organized "to show the world what lies in the path of destruction." He explained, "The Action Run will demonstrate that even the fastest runner will not be able to outrun the fury of GLOF. This devastation will not only cause loss of life and property but could also cause cultural and traditional values to fade and become extinct" (ICIMOD, 2012).

Earlier that year in March, 20 SYT members were invited to the ICIMOD office in Kathmandu for a half-day briefing and interaction program on climate change and its impact on the Himalayas. In this program, biodiversity in a changing climate, melting Himalayan glaciers, and the role of youth in climate change were discussed to raise awareness and knowledge of Sherpa students (ICIMOD, 2012).

In 2010, I attended the second "Beat the GLOF Action Run" program in Khumjung, a Khumbu village. This event started early in the morning with the runners prepared to race. The runners were mostly from Khumbu with a few of them from outside the region. After the run, people gathered in front of the stage placed in the Khumjung school playground near the entrance to the Khumjung village where rows of chairs were arranged. A few local Sherpa men, the organizers, media personnel, and international guests were invited to take the chairs in the front while the local residents stood outside the barred seating area. The seating arrangement showed clear separation of space that occurs between organizers and the local people during formal programs.

The program started with salutations to the sponsors and the international guests by the organizers. This was followed by Sherpa dances and individual speeches by the guests highlighting GLOFs as the climate change effects and the implications of climate change risks to a poor country like Nepal. The speeches centered on the potential Imja GLOF and the need for international attention to the Everest region. The last speaker at the program was a Sherpa man from Namche, who clearly received the loudest applause from the audience. He advocated for transparency by the researchers and for increased participation of and consultation with the Sherpas in climate change studies. He also requested the media to provide factual information about the tourism industry in the Everest region. Prior to this speech, national television shows had portrayed the region as extremely expensive and the Sherpas as wealthy hotel owners mistreating porters. There was a clear distinction in the content of the speeches given by the organizers, their guests, and the Sherpa man. While the organizers took a broader scale of discussing climate change – one that is prevalent at the national and international levels – the Sherpa man focused on the local issues that mattered to his fellow villagers. After the speeches, attendees of the event visited food stalls stocked with Sherpa delicacies run by local women's groups. The daylong event concluded with a dance party playing popular Nepali songs, which lasted until around 4 o'clock the next morning.

A year later, during an interview a Khumbu man in his 50s was critical of the organizers and questioned the effectiveness of this event in bringing long-term benefits to the people. He mentioned that the event benefitted the organizers more than the villagers. He did not, however, explain his reasons. His son quickly pointed out that although the event was not very effective, it was a small step toward achieving the larger goal of climate change adaptation through awareness. In 2014, while discussing climate change issues with new group of SYT members in Kathmandu, I found that they were also critical about the effectiveness of past SYT climate change events but were quick to point out other benefits these events provided such as opportunities for institutional networking and organizing community events.

Case IV: 2011 Andean-Himalayan Glacial Lake Exchange and Collaboration Expedition

On September 5, 2011, more than 30 international glaciologists and scientists left for Lukla for an 18-day field expedition and survey of the potentially dangerous Imja glacial lake in Khumbu. The expedition team represented physical and social scientists from Central Asia, Japan, North America, and Europe. This expedition was part of the "Promoting South-South Knowledge Exchange and Collaboration between Experts and Practitioners from the Hindu Kush-Himalaya, Andes, and Central Asian Mountain for Improved Adaptation and Resilient Livelihoods," supported by The Mountain Institute (TMI), USAID, National Science Foundation (NSF), Adaptation Partnership, University of Texas in Austin, ICIMOD and the United Nations Development Program (UNDP). The press release of the expedition described that throughout the nearly three weeks of field expedition the world-class climate change, glacial lake, and adaptation specialists would exchange experiences and identify priority needs in climate change action research and pilot projects. The goal of this expedition was to develop strategies for increasing exchange and collaboration opportunities for practitioners from their home mountain ranges – the Andes, Tien Shan, Pamirs, Karakoram, and Hindu Kush-Himalaya.

This expedition used an online blog, updated daily to share their experiences "to allow people around the world to follow and learn about the challenges of working in this remote mountain region, and to share challenges faced by local mountain people who feel the impacts of climate change first." With the followers of the blog being able to ask questions to the expedition members in real-time, it was considered to be a "groundbreaking and fascinating" blog to follow (The Mountain Institute Blog, 2012).

The first Andean-Asian Glacial Lake Expedition to the Everest Region was expected to provide an opportunity for the Peruvian specialists with more than 50 years of engagement in the control and management of more than 35 dangerous lakes in the Cordillera Blanca to share their experience with participants from Central Asia and the Hindu Kush-Himalayas, where the phenomena of new glacial lake development is more recent. During the expedition, members spent three days camped below the Imja glacial lake to exchange experiences in glacial lake control, mitigation, and management, with the participation of local people.

The Sherpa villagers in Khumbu hosted the scientists in their lodges. Several meetings with the Sherpas were also conducted in Khumbu. A year later, in an interview with a BBC journalist, a Sherpa man from the same Khumbu village where the expedition team conducted their meetings, said, "Every time we begin to forget about the threats from glacial lake outburst, then comes news of yet another study through the radio and television, and this has been happening over and over again for more than 15 years now … Instead of having to fear death like that again and again, we would rather die once if the lake really bursts out one day" (Khadka, 2012).

When the scientists returned to Kathmandu after the expedition, a workshop was conducted to identify priority needs for climate change action research and pilot projects focusing on the potential Imja GLOF as the climate change effect. Noticing that there were no Sherpa residents from the region at the workshop while the priorities for climate change action research were being listed, I pointed out to a group of scientists tasked to focus on climate change impacts on the people that an absence of Sherpas in such workshops can lead to difficulty in implementing programs in the villages later. Then, I asked how this group would plan on ensuring participation of the Sherpas in future climate change programs. To this, an international glaciologist replied, "Of course we are going to make sure the Sherpas are participating, they can carry the pipes to Imja."

On September 17, 2011, an earthquake with a preliminary magnitude of 6.9 was felt across northeast India and Nepal. Conversations in the villages after this earthquake further revealed the distance between the international scientists and the local villagers. In the Everest region, although no one was seriously injured, the quake damaged newly built houses, bridges, and pathways. It left the Sherpas in the Everest region panic stricken. When asked about the experience the next day in Pharak, informants from three villages said they were terrified and did not know what had caused the earthquake. One of my informants said that she thought the earthquake was a part of the apocalypse portrayed in the Hollywood movie *2012*, which is based on the end-of-the-world interpretation of the Mayan calendar. Another informant, whose house had been badly damaged by the quake, said she thought Imja Lake had flooded. She further added that she was scared about Imja Lake because a group of international scientists were on an expedition to the Lake, who probably came because Imja was about to burst, "why would they come if that was not the case?"

CONCLUSION

As Cases I, III, and IV show institutional climate change adaptation efforts cannot assume that by reaching out to a few individuals in the region they will benefit the whole. If the current approach of reaching out to the same group – middle aged, male hotel owners from on-route villages – is continued, institutions will continue to expand and maintain scientific knowledge gap among Sherpa individuals, and the invaluable firsthand environmental change knowledge from farmers and herders will continue to be ignored and marginalized (Sherpa, 2014). Similarly, communities at the local level cannot be assumed to be homogeneous entities.

Social heterogeneity even when working with a small-scale cultural world should be considered and understood as a dynamic characteristic. Social heterogeneity as I discuss here for Pharak Sherpas through age group, gender, occupation and residence, for example, is not fixed or static. Due to the dynamic nature of the society and the economy, social heterogeneity is inevitably dynamic.

Organizing institutional climate change activities to draw international attention alone as in Case II is not sufficient to address climate change adaptation issues. While the magnificent mountains and the Sherpa culture could draw international attention and resources, those attractions alone are not enough. Institutional climate change adaptation efforts should go beyond that and involve local residents as equal partners and not just as passive recipients if effective solutions are expected.

Recognizing climate change adaptation as a pivotal concern, in order to practically, effectively and sustainably address the challenges I argue that we have to expand our focus from just the Sherpa community to the Sherpa cultural world, where institutions are also participants. It is crucial that not only the Sherpa agency (or lack of agency) or understandings are examined but the institutional engagements and delivery are also assessed. While the vital role of institutions cannot be dismissed, providing scientific knowledge and resources, the firsthand experiences, observations, and the multiple knowledge systems that exist at the local level also cannot be dismissed as scientifically inadequate.

Ethnographic assessment of institutional climate change adaptation efforts among Sherpas of the Mount Everest region shows the human dimensions of climate change. It shows the complexity in working on adaptation issues at the local level. On the one hand, institutional efforts have become a source of rumor putting lives in danger as Case I shows. On the other hand, institutional efforts also provide knowledge and resources, although unevenly, to the local residents as Cases I and III show. Institutional activities have increased local receptivity to scientific climate change knowledge, but it has also increased fear of an impending doom, and anger over the continuous discussion of climate change without concrete actions.

ACKNOWLEDGMENTS

I would like to acknowledge the financial support received from the Wenner-Gren Dissertation Fieldwork Grant, the Harka Gurung Research

Fellowship through Social Inclusion Research Fund in Nepal, and a research fellowship from the Association of Nepal and Himalayan Studies. Their generosity assisted in timely completion of my research. I would like to thank John Bodley, Barbara Brower, Nancy McKee, Late Pralad Yonzon, and Ram Chhetri for their mentorship throughout my research. This paper benefitted from the valuable comments by Jim Fisher, two anonymous reviewers and Donald Wood, editor for Research in Economic Anthropology. I take full responsibility for any errors that remain. Author is a Cultural Anthropology Lecturer in the Department of Anthropology at the Pennsylvania State University.

REFERENCES

Adger, W. N., Lorenzoni, I., & O' Brien, K. (2009). *Adapting to climate change: Thresholds, values, governance*. Cambridge: Cambridge University Press.

BBC. (2009). *Nepal cabinet holds meeting on Mount Everest*. Retrieved from http://news.bbc.co.uk/2/hi/south_asia/8394452.stm

Bodley, J. (2003). *The power of scale: A global history approach*. New York, NY: M.E. Sharpe, Inc.

Bodley, J. (2012). *Anthropology and contemporary human problems*. Lanham: AltaMira Press.

Brower, B. (1991). *Sherpa of Khumbu: People, livestock, and landscape*. Delhi: Oxford University Press.

Byers, A. (2007). An assessment of contemporary glacier fluctuations in Nepal's Khumbu Himal using repeat photography. *Himalayan Journal of Sciences*, *4*(6), 21–26.

Fisher, J. (1990). *Sherpas: Reflections on change in Himalayan Nepal*. Berkeley, CA: University of California Press.

Government of Nepal. (2014). *National population and housing census 2011*. Solukhumbu: Central Bureau of Statistics. Retrieved from http://cbs.gov.np/wpcontent/uploads/2014/04/11%20Solukhumbu_VDCLevelReport.pdf

High Mountains Adaptation Partnership. (2014). *Khumbu LAPA complete reference document*. Retrieved from http://www.highmountains.org/content/khumbu-lapa-complete-reference-document

ICIMOD. (2012). *Information resources*. Retrieved from http://geoportal.icimod.org/NAE/NewsDetail.aspx?NewsID=132

Khadka, N. (2012). Everest Sherpas in glacial lake study warning. *BBC News*. Retrieved from http://www.bbc.co.uk/news/science-environment-19569256

Ministry of Environment. (2010). *National adaptation programme of action to climate change*. Singha Durbar, Kathmandu, Nepal: Government of Nepal, Ministry of Environment.

Ministry of Environment. (2011). *Climate change policy. Nepal climate change and development portal*. Retrieved from http://www.climatenepal.org.np/main/?p=research&sp=onlinelibrary&opt=detail&id=419

Ortner, S. (1989). *High religion: A cultural and political history of Sherpa Buddhism*. Princeton, NJ: Princeton University Press.

Sherpa, M. (2013). *Conservation governance and management of Sagarmatha (Mt. Everest) National Park, Buffer Zone, and Buffer Zone community forest user groups in Pharak, Nepal.* PhD Dissertation, University of Massachusetts at Amherst.

Sherpa, P. Y. (2014). Climate change, perceptions and social heterogeneity in Pharak, Mt. Everest Region of Nepal. *Human Organization, 73*(2), 153–161.

Sherpa Kyidug. (2012). *Sherpa Kyidug.* Retrieved from http://www.sherpakyidug.org/

Shrestha, A., & Aryal, R. (2011). Climate change in Nepal and its impact on Himalayan Glaciers. *Regional Environmental Change, 11*(1), 65–77.

Skog, L. A. (2010). *Beyul Khumbu: Sherpa constructions of a sacred landscape.* MA thesis, Portland State University.

Somos-Valenzuela, M., McKinney, D., Byers, A., & Rounce, D. (2013). *Bathymetric survey of Imja Lake, Nepal in 2012.* Austin, TX: University of Texas, Center for Research in Water Resources.

Spoon, J. (2011). The heterogeneity of Khumbu Sherpa ecological knowledge and understanding in Sagarmatha (Mount Everest) National Park and Buffer Zone, Nepal. *Human Ecology, 39*(5), 657–672.

Spoon, J., & Sherpa, L. N. (2008). Beyul Khumbu: The Sherpa and Sagarmatha (Mount Everest) National Park and Buffer Zone, Nepal. In J.-M. Mallarach (Ed.), *Protected Landscapes and Cultural and Spiritual Values* (pp. 68–79). Gland, Geneva, Switzerland: World Conservation Union.

The Mountain Institute Blog. (2012). *Expedition press release.* Retrieved from http://www.mountain.org/blog/2011/09/the-mountain-institute-and-the-imja-lake-expedition-team-says-thanks/

Von Furer-Haimendorf, C. (1964). *The Sherpas of Nepal: Buddhist highlanders.* London: J. Murray.

Watanabe, T., Kameyama, S., & Sato, T. (1995). Imja glacier dead-ice melt rates and changes in a supra-glacial Lake, 1989–1994, Khumbu Himal, Nepal: Danger of lake drainage. *Mountain Research and Development, 15*(4), 293–300.

Waylen, K. A., Fischer, A., McGowan, P. J., & Milner-Gulland, E. J. (2013). Deconstructing community for conservation: Why simple assumptions are not sufficient. *Human Ecology, 41*(4), 575–585.

INSTALLING ROOFTOP SOLAR PANELS IN PRIVATE HOMES: THE CASE OF SMALL ELECTRICITY PRODUCERS ON REUNION ISLAND (INDIAN OCEAN)

Sophie Bouly de Lesdain

ABSTRACT

Purpose – *In France, as in other countries, the idea of installing rooftop photovoltaic (PV) panels in private homes is based on an incentive scheme (tax advantages, feed-in tariffs, etc.) inspired by neoclassical economic theory. In the case of electricity producers in Reunion Island, unlike economists, we argue that producers' calculations involve decision-making criteria which go further than any simple evaluation of economic costs and benefits.*

Methodology/approach – *Our approach is based on concepts of economic anthropology and on observations and semi-structured interviews conducted in the homes of the producers.*

Climate Change, Culture, and Economics: Anthropological Investigations
Research in Economic Anthropology, Volume 35, 25–46

ISSN: 0190-1281/doi:10.1108/S0190-128120150000035002

Findings – *This ethnographic method allowed us to examine economic rationalities which revealed the anticipation of an energy landscape that will be subject to issues relating to the environment, access to electricity, evolution in the local electricity market, and household budget management. In this context, producers' representations of solar power and of processes for commoditizing and decommoditizing the electricity produced (sold on the network/"free" when consumed) make compatible preservation of the environment and social norms of consumption.*

Implications – *This paper focuses on PV energy producers (who have been the object of very little research) and thus provides input for existing reflection on the diversity of economic rationalities. Such insight is important for understanding how people respond to policy appeals for PV panels. Anthropology therefore has an important role to play in the debate on energy transition. This conclusion paves the way for similar research in other contexts (of a non-insular nature in particular) which would allow for a promising comparative anthropological approach.*

Keywords: Research paper; small electricity producers; "ordinary practices of calculation"; Reunion Island; electricity production and consumption

INTRODUCTION

Like other French Overseas Departments, Reunion Island (Indian Ocean) has the particularity of being dependent on fossil fuels – which are imported for electricity generation and are responsible for environmental damage – and not on nuclear power as is the case for mainland France. The current energy and environment policy is designed to help manage the demand and to increase the proportion of renewable energies used.

Among the battery of measures which have been introduced, households with photovoltaic (PV) panels have the particularity of allowing people to both use and sell the electricity they produce. In terms of environmental policy, the installation of PV panels in private homes pays "Double-dividende" (Keirstead, 2007): it reduces recourse to fossil fuels and may cause private producers to moderate their electricity consumption. While Reunion islanders who become small-scale energy producers benefit from tax advantages and a regional subsidy, and sell their electricity at an attractive price, the electricity they buy remains one of the cheapest in the EU.[1]

In France, as in other countries, environmental policies are based on incentives (feed-in tariffs, tax advantages, etc.) which encourage resale of the electricity produced, rather than personal consumption. The environmental policy is therefore based on a unique conception of economic rationality inspired by neoclassical economic theory (as far as maximizing profit is concerned).

Most social scientists, including anthropologists, have criticized the behavioral approach which is dominant in public policies dedicated to the environment, both in France and abroad. This approach postulates that individual choices are dictated by the economy or technology (Lutzenhiser, 1992; Wilhite, Shove, & Lutzenhiser, 2000)[2] and only concern end users (Wilhite, 2008). A field of anthropological research has thus grown up around these issues. "The anthropology of electricity" aims to define the cultural and social dimension of electricity practices and of energy practices in general. There are no "energy practices" as such, but rather social and cultural practices such as washing, cooking, and so on, which involve energy being used (Shove, 2003), and which fall within the fields of anthropology. Some anthropologists have chosen comfort as their preferred area of research (Chappells & Shove, 2005; Shove et al., 2008), sometimes adopting a comparative approach so as to demonstrate the cultural aspect of such practices (see Wilhite et al.'s works comparing Japan and the United States, 2001) or the impact that electrification might have on countries in the southern hemisphere (Strauss, Rupp, & Love, 2013; Winther, 2008). This research concerns both the consumption and production of energy. In Sweden, Annette Henning (2000, 2005, 2008) has underlined the cultural dimensions of energy practices (e.g., gender, power), which in turn explain recourse to energies (e.g., solar or biofuel) or systems (e.g., pellet-solar system) which are more or less environmentally friendly. Her approach echoes sociotechnical analysis. In her seminal article "How to describe technical objects," Akrich (1987) uses the example of a "solar kit" funded by an NGO to demonstrate how users can shape technology – and in turn be defined and transformed by technology. The author talks about the processes of "co-construction" between techniques and their users (and designers) which lead us to take a particular interest in contexts for the adoption of technical devices – such as solar panels. In this respect, Knight and Bell (2013) use the country's historical context (and memories of the 1941–1943 famines in particular) to explain the development of PV panels in Greece after 2007. PV electricity producers tend to be people who are suffering from the economic crisis and who are looking for additional income. However, the authors do not provide any details of the calculations that these producers made when coming to their decision to install solar panels.

In the case of electricity producers in Reunion Island, unlike economists, we argue that their calculations involve decision-making criteria which go further than any simple evaluation of economic costs and benefits. These calculations relate to what Wéber (2010, p. 367) calls "ordinary practices of calculation," that is, "modes of assessing costs and benefits, monetary or otherwise, at work in the decisions (everyday or exceptional, conscious or automatic) that individuals make (...)."

This paper focuses on the calculations these electricity producers make when purchasing PV panels: to what extent do economic incentives affect their decisions? To what extent do they take environmental concerns into consideration? Are these two aspects (economic and environmental) contradictory? Do they affect people's relationships with electricity and the ways they use it? The latter question will require us to consider the status of the electricity that is produced (solar power from your own rooftop PV panels *versus* electricity from the grid; resold merchandise/self-produced commodity) and its impacts on the electricity consumption of the domestic producers we met.

Before we move on to answer these questions, let us take a look at the local electricity system (including demand and means of production), at the electricity market and at the objectives of Reunion Island's environmental and energy policies. This will give us a better understanding of the context in which the interviewees are living. We will then present our methodology and the interviewees (age groups, activities, etc.). Having explained the basis for our study, we will take a closer look at how environmental issues and financial calculations (formulated in terms of methods of financing, budget management, etc.) affect the decision of whether or not to install PV panels. Lastly, we consider the link between the decentralized electricity production and consumption of producing households.

CONTEXTUAL ELEMENTS: THE ENERGY SYSTEM, THE ELECTRICITY MARKET, AND ENVIRONMENTAL POLICY

Located in the Indian Ocean, Reunion Island has a tropical climate, with oceanic overtones; the further inland you travel, the higher the altitude, with the coastal oceanic climate becoming a tropical mountain climate.

Demand for electricity constantly increased over recent years. With 833,000 inhabitants[3] it is the most populous French Overseas Department. From 1970 to 2010, Reunion's population virtually doubled, from 450,000 to just over 828,000. To this demographic growth must be added an improved standard of living in the form of greater comfort and better equipment in homes. Since 1990, comfort has been especially improved through the generalization of sanitary facilities. Hot water has been the fastest spreading facility: only 38% of Reunion homes had hot water in 1990, 63% in 1999, and 82% in 2008.[4] This access to consumption (cooker, refrigerator, television, hot water, etc.) requires electricity. Demographic growth and catching up with mainland France in terms of household appliances thus led to a 205% increase in household electricity consumption between 1980 and 2005 (Savidan, Shaffar, Dimou, & Garde, 2008). The island's population is likely to reach one million by 2030 and the average electricity consumption of a Reunion family has not yet reached that of households on the mainland.

The electrical system has to meet this increasing demand while at the same time conforming to environmental policy requirements. Reunion Island, the other overseas departments, and Corsica differ from mainland France's dominant electricity system in that they have no nuclear power; on the other hand, they import and use fossil fuel to generate electricity. This was not always the case for Reunion Island. In the 1980s, hydraulic power provided 100% of the island's electricity, but by 2000 this figure had fallen to just 32%. Over the same period, coal and oil increased from 0% to 21% and 19% of the energy mix, respectively.[4] Renewable energies (hydraulic, bagasse, solar panels, wind power, biogas) now represent almost 35% of the energy mix, the remaining 65% being made up of fossil fuels (coal, heavy fuel oil, and diesel). Finally, since 2007, the French electricity market has been open to competition – with the exception of these overseas territories due to the high production costs. These costs are not passed on to the islanders, who benefit from "tariff equity" – by virtue of the principle of territorial equity, electricity is sold at the same price throughout all territories of the French state. Electricity is therefore considered to be a right to which every citizen must be allowed access. Electricité de France (EDF) is responsible for the production, distribution, and marketing of electricity throughout the island. It also has the legal obligation to purchase electricity supplied by other producers. So the state very much puts its stamp on the electricity market and sets regulated tariffs for both purchase and sale. Finally, just like the rest of France, electricity remains one of the cheapest in Europe.

As far as national environmental policy is concerned, Reunion Island (and, more generally, all other overseas departments) offers optimal conditions for energy transition. Indeed, the island imports fossil fuels – at a high cost – and has natural resources which are very favorable to renewable energies. Organizations have been created to implement the island's energy policy. The region created the Agence Régionale de l'Energie Reunion (ARER) in 1998, followed by PRERURE, the Observatoire de l'énergie, GERRI (the regional version of the Grenelle environment forum) and, finally, Energies Reunion in 2013. The deadline for the island's energy autonomy was set for 2025, and then pushed back to 2030. This will notably involve improving demand-side management in the residential sector, which represents 40% of final electricity consumption (43.5% in 2007).[5] Citizens are also being targeted in relation to the development of renewable energies. They are being encouraged to fit their homes with solar panels and to sell the electricity they produce – feed-in tariffs favor small units (less than 9 kWp) for private homes. There is thus a shift from centralized production requiring major power plants (hydraulic, oil, etc.) to a decentralized system where residents have their own solar energy production unit. This offers Reunion islanders a new role; giving them the opportunity to play a more active part in bringing renewable energies to the island.

RESEARCH DESIGN AND METHODS

In this context, we conducted semi-directive interviews (approximatively 30) in producers' homes, in order to see their houses, their environment, and their rooftop PV panels. All these interviews were recorded and transcribed. They lasted between 50 and 90 minutes. Several members of the household were sometimes present during the interviews. Our ethnographic method allowed us to examine profitability calculations (if any), energy management, interviewees' notions of the electricity market and its evolution, their relationship with the island and with the environment, etc. These subjects were discussed to varying extents, with a variable amount of time being given to the economic and environmental issues.

All of the producers we met live on the west coast of Reunion Island, own their homes, and installed their solar panels after the fall in feed-in tariffs in 2010,[6] so they had not been producing for long; some were even in the process of installing their panels at the time of the interview. They own rooftop PV systems of less than 3 kWhp, connected to the grid and to a battery. They sell all the electricity they produce. When talking about their installations, they use the terms "photovoltaic panels" (PV panels) and "solar energy"; we will therefore use the same terms in this paper.

They belong to the upper and middle classes and are teachers, farmers, nurses, architects, retired civil servants, etc., which ensures them a comfortable income in an economic context where the level of unemployment stands at almost 30% of the active population (compared to 10% in mainland France). PVs are less accessible to poorer families because of the price of PV equipment. After deducting tax and regional aid,[7] the break-even point is estimated at 10 or 12 years, on the basis of an EDF feed-in tariff of 20 years at 0.3266€/kWh TTC in July 2013, and pay 0.1329€ for the electricity they consume. Even though it has fallen, this tariff does not favor personal consumption of the electricity produced.

PARTICIPATING IN THE ISLAND'S FOSSIL FUEL INDEPENDENCE

In such a context, some of the producers we met focused on environmental issues. These producers clearly expressed concerns about the fossil fuels available on the island, and therefore about the environment. Xavier illustrates this case. Xavier is a retired civil servant, and like other small-scale producers he wants to "consume locally"; for him, this means being able to use forms of fuels he refers to as "clean" and "free" (sun, water, etc.) as opposed to "expensive" and "dirty" (coal and oil products). Yet he sees no difference between directly using the electricity he produces, and feeding it into the network, because: "*Electricity is electricity. It's not because I'm the one producing it that it's better. It's not as if I were planting carrots or buying local carrots. The physical principle is the same, the current is the same.*" For Xavier, whether he uses it or not, his production is helping to reduce the proportion of fossil fuels used on the island and to increase that of "green" energies. He spoke generally about "the environment" and how fossil fuels damage the island. His house has its own source of water and the materials used to build it were chosen so as to reduce his carbon footprint. There is no need for air-conditioning because his house "*is bioclimatic, there's a corridor running through it, in the shape of a cross, so air comes in from all sides, it's naturally air-conditioned. The partition walls also provide ventilation because they're made of wood, they're not closed off at the top, so the air circulates inside.*" For Xavier, his house is part of a process of autonomization – true autonomy being a far-off objective – which he had in mind when he designed the house. "*When I made the plans for the house, I integrated the need to include renewable sources of energy and water collection. I proceeded in stages, the water tank first, then the solar-powered boiler, then the photovoltaic panels.*" Like other producers, the purpose of Xavier's solar panels is to achieve energy autonomy and his low consumption of electricity is already helping to achieve this, which is in line with the Region's objective of reaching energy autonomy by 2030. This position is not unique and some producers, like Valérie, chose to install solar panels specifically as part of the regional agenda: "*The year 2030 is very much in everyone's minds, young people are talking about it a lot.*"

Some of these producers see themselves as "ecologists," but this does not mean they feel they have to adopt certain practices (such as dry toilets) or reject the consumer society, as this corresponds neither to their lifestyle or their aspirations. So the desire for autonomy does not extend to all areas of everyday consumption (not just energy and water, but also food, health, etc.) and does not involve the militant lifestyles that can be found elsewhere. Geneviève Pruvost (2013) describes a set of personal production practices which are accompanied by relationships with work (excluding personal production) and with consumption that lead to "radical lifestyles" which are far removed from what we observed in Reunion. These islanders nevertheless share with activists the notion of autonomy (albeit limited to just electricity and water) as a process. The installation of solar panels is a first step and is intended to evolve (additional panels) to meet all electricity requirements, even those which do not yet exist, such as heating a swimming pool.

These producers did not mention money. And when I raised the question of the profitability of the PV panels with Madeleine, she explained that she had chosen that particular installation because "*the other offers always talked about money, saying that 'you will make money' – the arguments were too much about economics.*" These producers claim that they never made any profitability calculations. They paid cash for their PV panels, or took out a loan that will be repaid by the feed-in tariffs, and they talk about a "cash-neutral" operation with neither profit nor loss. EDF's feed-in tariffs created a renewable energy market (Debourdeau, 2011) from which these producers seem to wish to disengage. Talking about profit does not appear to fit in with the environmental considerations behind the installation of PV panels. At the very most they talked about a "balance" between ecological necessity and financial interest, as is indicated in this quote: "*How can I put it, Reunion can't consume an infinite amount of oil; I thought it was interesting to take part in some way in an ecological project, and there was a very concrete benefit (...)... in a sense we made an ecological decision and we balanced the operation financially, so that's pretty much it, to sum it up*" (Marc). The situation is the exact opposite of the one that Annette Henning observed in Sweden (2008). The author underlines the social valorization of economic arguments, which explains why people use these to justify their energy choices for heating.

For the interviewees mentioned so far, the point of producing electricity is to avoid being a burden on the island's consumption as a whole, and to limit the use of high-carbon fuels which cause environmental harm. These producers are not part of any political protest or breakaway movement (production in this case) – such as the "off-grids" in the United States. For the latter, installing your own rooftop PV panels is a "deliberate choice to be less dependent on the State and on the 'system'."[8] In France, prior to 2000, when EDF became obliged to buy the electricity which is produced, the decision to install PV panels was either an alternative to an expensive connection to the power grid, or else a militant act which was both pro-environment and anti-nuclear (for mainland France) or anti-fossil fuel (for overseas islands). Even nowadays, decentralized production provides electricity to homes located in certain isolated parts of Reunion Island. These homeowners have no possibility of being connected to the grid. The situation is very different in the case we are examining here. Some of the producers we met have taken the decision not to be dependent on the grid; but this is a decision to no longer have to put up with power outages.

PRODUCING TO ACHIEVE UNINTERRUPTED SUPPLY: POWER OUTAGES

This forces us to look at private electricity production from a different angle: the objective of installing solar panels is thus to reduce dependency on the grid so as to maintain uninterrupted consumption, and not to reduce the impact of personal consumption on the production of fossil fuels or to supply the island with green energies.

Reunion Island is regularly subject to power outages lasting for several hours, or even several days. At the time of writing, the most recent major power outages date back to the winter of 2013 (between the end of December 2013 and the beginning of January 2014), following Hurricane Bejisa. To this we can add frequent micro-cuts which deteriorate the quality of electrical appliances in general and computer equipment in particular, and the load-shedding carried out by EDF to manage production. By definition, and with the possible exception of an interconnection such as that between Corsica and Sardinia, insular territories cannot turn to neighboring grids when there is a production shortage. The geographical contours of these territories define the electricity grid. Islanders are de facto aware of energy issues, through the absence or interruption of the electricity supply, and thus of certain daily activities.

As we have seen above, the interviewees sell their entire production – they are therefore billed for their consumption. But when there are power outages, they are able to use their own electricity, including that which is stored in their batteries. In this case, they have to pay particular attention to their electricity consumption in relation to the amount available (in the battery or supplied by panels). When there are lengthy power outages, the interviewees prioritize cold storage (freezers and refrigerators). Even when one has a battery, this does not remove the possibility of mishaps, especially if the said battery does not always work properly – so when it is hot, the contents of the fridge can end up in the bin. Pierre had such an unfortunate experience during Hurricane Bejisa: "*We'd gone to France (...) we thought it was all right to leave everything [because there was a battery]... so I just left one thing running .... [the freezer] .... We had to throw everything out, we had to do the shopping all over again, when we were supposed to be going back to work ....*" The second priority is cell phones, in order to keep in touch with friends and family. It is vital to have a link to the outside world when there is a hurricane, as injuries and/or material damage can be serious. During Hurricane Bejisa, Jacques used some of his own electricity but also gave some to his neighbors:

> Our light was a source of envy! The entire district was blacked out ... no electricity for a week, no telephone for a month. Everyone's cell phones were dead, so I had rows of phones belonging to people in the district! The neighbour who lives opposite brought me his fish, his yoghurts in the second fridge! It's good for that too, it creates links with your neighbours – there were already some links but it helps to strengthen them, by helping each other. (Jacques)

Thus social links are also maintained within smaller geographical perimeters, particularly with close neighbors. Producing electricity is no longer about creating a link with the territory as a whole, in relation to a common objective (reducing the use of fossil fuels and thus reducing carbon emissions), as in the previous case, but about creating a link with a smaller environment (the neighborhood). This creates a territory which is divided into zones defined by access to electricity, and which in Reunion Island separates "les Hauts" (the mountainous regions) from "les Bas" (the coastal regions). In the case of power outages, neighborhoods are reconstructed through exchange and solidarity. Personal production allows people to access and help supply an exchange network (in relation to workers, see Pinçon, 1986; Wéber, 2009). These solidarity networks help to build a common sense of belonging in relation to access to electricity and to memories of extreme weather events which plunged the island into darkness and played havoc with everyday life.

The mountainous regions of the island are particularly vulnerable to power outages. Islanders who have moved to zones with a better service, or who live along the coast, still remember power outages dating back several decades. To explain his decision to install solar panels, Jacques talked about a 1980s hurricane, as if it had happened yesterday, which had deprived him and his family of electricity for 45 days. Given that one needs access to electricity in order to also have access to water: "*Well, in Saint Leu I waited for 45 days. I didn't have any water either, the water pumps weren't working. That was in the 1980s, the children needed milk, yoghurts, fresh food.*" Families with children, as was his case back then, spoke more than others about the problems caused by power outages which lasted several days. Their everyday lives are regulated by the needs (meals, baths) of their children and are therefore less flexible. In such a context, the "emergency" function of the equipment (the battery) provides reserve electricity and thus a certain consistency in the level of comfort.

Lengthy power outages aside, micro-cuts are frequent and this damages equipment, especially computers.

In most cases, the source of the electricity (solar-generated) does not explain the decision to install PV panels. What is important is to have electricity so as to have access to a certain number of domestic appliances. This is by no means self-evident in Reunion Island. The interviewees accepted that power outages have become less frequent over recent years; but some of them have a relatively pessimistic vision of the island's electricity in years to come. For Georgette and other producers, there is no doubt that there are natural geographical limits (the contours of the island) to local production and that the electricity system will not be able to meet a demand that is constantly increasing in line with the population: "*we are going to reach a population of one million, so there is bound to be a shortage of electricity*." From this perspective, by producing her own electricity she will have access to electricity in the future and will not have to rely on an overloaded grid.

Like others, Raoul hopes that his home will be self-sufficient in the future, using his low level of electricity consumption as an argument. Yet at the same time he admits that he does not know how much electricity he uses – he assesses his consumption by looking at what he has to pay ("*I couldn't tell you how much we use and I haven't even had a look at the bills ... all I know is that we don't use much because we don't have air-conditioning.*" – Raoul).

The issue of access to electricity is sometimes raised in financial terms rather than in terms of power outages. In both cases, owning your own rooftop PV panels maintains access to a level of comfort which is relatively new to some of the interviewees. Such calculations put solar production into an economic context, which as we have seen, was not mentioned in the earlier discourses. They lead us to examine how producers assess the profitability of their units.

YOU INSTALL PV PANELS TO "*SAVE MONEY,*" TO BENEFIT FROM THE SUBSIDIES, AND TO REDUCE YOUR BILLS

As far as the producers are concerned, purchasing PV panels means "saving money" because taxation, regional aid, and the feed-in tariffs partially reimburse the cost of the installation ("*you break even after 15 years,*" – Victor). So, here, the cost of the installation is not assessed as a whole (including loans, where relevant), but rather in terms of how the monthly payments affect the household budget. "*If it was for 3 months, or for 2 or 3 years, I wouldn't have done it; but this is for the long term, the cost of the PV won't affect my budget too much* (...)," said Jeanin, showing that the cost of the installation (whether it is considered to be expensive or not) is estimated on the basis of constraints which are often ignored by classic economic theories.

> As far as the investment was concerned, it was much cheaper to invest immediately, it was much easier to accept, because in the long term I saw that the investment was cost-effective, plus it didn't have any major effect on my standard of living. (Madeleine)

There was an immediate comparison with solar-powered water heaters; the region used to provide financial aid for their installation. Since 2009 they have been obligatory in all new homes,[9] but are no longer subsidized. The number of homes on the island equipped with solar-powered water heaters thus rose by 430% between 1999 and 2008.[10] Almost 50% of private homes are now equipped – a proportion which reaches 80% in Tampon and 71% in Saint-Pierre, where the majority of the interviewees live. They believe that home-produced electricity will reach the same levels. So interviewees are having them installed while they can still benefit from financial aid and from an attractive feed-in tariff.

To Save Money, or Maybe Even to See the End of Electricity Bills

This reference to solar-powered water heaters also explains the attraction of using home-produced electricity, even though the feed-in tariffs favor resale. In people's minds, installing PV panels today means less expense tomorrow (cost of the PV panels excluding aid), and not having to pay electricity bills, with the interviewees hoping that they will have less or maybe even nothing to pay: "*to begin with I thought it would bring the bills down a bit, so I'd save electricity*" (Jacques). The expenditure is seen as a saving. This is in line with a form of "money earmarking" analyzed by Zélizer (1994). She shows how money is not a neutral vector of exchange but is socially charged: money serves to meet different expenses, depending on whether it comes from a salary, for example, or from a family allowance. In this case, some interviewees anticipate that the income generated by the sale of electricity will pay their electricity bills. So the future profit from selling on electricity is not seen as potential income, but as a reduction of expenditure related to the electricity bill: "*Bah, I don't know, as far as I'm concerned it's not really a source of income, it's just that I'll spend less*" (Valérie). In a similar vein, Germaine, who is an old-age pensioner, decided to install solar panels because her bills had significantly increased and she did not understand why. For her, as for others, electricity was a technical field reserved for engineers, totally foreign to laymen. So she chose to eliminate the problem (bigger bills that she could not explain) by producing electricity, to balance her increased consumption.

The reason for installing solar panels might therefore be to eliminate not just the cost of the bills, but also their physical form: some small-scale producers do not understand why they need to bill EDF for the electricity they sell on, and at the same time be billed by EDF for the electricity they use. They think that EDF "should simply" deduct what they use from what they produce. In the following quote, Philippe clearly expresses this vision: "*The question is this – instead of receiving a payment from EDF, can't it just be deducted from the following year's bills? The system has evolved. Let's imagine I make monthly payments in advance and then at some point my consumption plummets – but I'm in credit. So ... either they pay me back the credit, or else they deduct it from the following year's bills, either way I have a credit with EDF because I've supplied electricity, it's like I've used less electricity ...*" (Philippe).

So these producers integrate their electricity production into their own way of thinking, which remains dominated by consumption. They think in terms of their overall "electricity budget," which includes both "income" (selling PV electricity) and "expenditure" (electricity bills). In her analysis of family accounting norms, Perrin-Heredia (2011) concludes that there is a hierarchy between fixed costs and variable costs which does "not depend on a criterion of 'absolute necessity' (what one needs to live) but on a criterion of 'contractualisation' (expenditure for which one has made a monthly commitment: rent, telephone, insurance, taxes, loans, etc.)." She thus observes that a calculation by subtraction (how much money remains to live on) from total expenditure operates alongside a calculation by addition (fixed costs). An electricity bill is a fixed cost (Winther & Bouly de Lesdain, 2013), which if it is reduced means there is more left over to live on. Marie installed PV panels to save money, and according to her, wealthy Reunion islanders are not interested in purchasing PV panels since feeds-in tariffs have fallen. The few producers from the upper levels of society who do install solar panels, would do so "pour la dentelle" – an expression meaning "for the prestige." for example, the "Conspicuous consumption" identified by Veblen (1899). For Marie, in this case, there is no question of budget management – just ostentatious consumption because the PV panels serve no material purpose. Like other producers we met, Marie puts PV panels in the category of middle-class objects.

Like Marie, most of the interviewees who focused on their electricity bills, hoped to use the electricity they produce ("*Because what I wanted was to use the electricity I produced*"), even though the feed-in tariffs are financially advantageous. Yet consuming the electricity they produce would remove their bill. These producers therefore face constraints (e.g., budget, power management) and have priorities which explain their choices but which run against standard economic logic.

Anticipating Possible Increases in the Price of Electricity

The desire to use the electricity one produces can be projected 20 years into the future, to the moment that the feed-in tariff ends. The French government's announcements about forthcoming increases in electricity prices have strengthened producers' faith in their calculations. As indicated in this quote, some producers wonder how the electricity market will evolve, and worry that the cost of electricity in French Overseas Departments will be called into question: "*Who's to say that the State will continue to allow Reunion islanders, whatever the electoral potential that might represent, to pay 30% less tax, to have the sun, no heating to pay for* ...." The decision to install PV panels is thus one way of guarding against any such changes.

Some interviewees fear that these increases, which will affect all of France, will mean the end of the special status accorded to overseas departments (price of petrol, tax rates, tariff equalization). The State's support for local consumption goes hand in hand with measures which support the feeling that Reunion Island (and French Overseas Departments in general) is part of a "separate" market ("*I think I'm right in saying that a few years ago it was about 40% cheaper. I don't know if it's because of the aid, the subsidies and all that, but here in Reunion Island we say that we're 'totally French yet totally separate'*" – Jean). This is not always experienced as something positive. For example, the expiry dates for foodstuffs are longer than those in mainland France ("*There's a whole bunch of laws and regulations which apply in mainland France but not here. Take use-by dates, they're six times longer! We're fully French, but on the mainland they eat their yoghurts within 2 weeks, here it's 6 months!*" – Jean). This feeling of being part of a separate market affects the views of some interviewees regarding the quality of the solar panels they bought ("*I was wondering what generation of panels we are getting here in Reunion, whether they are panels which are no longer made on the mainland. I'm wondering because here in Reunion Island there are examples in loads of different fields where we are sent stuff that is no longer used in France!*" – Marc).

In this changing context, the installation of solar panels guarantees price stability: for the interviewees, feed-in tariffs will compensate for any increase in electricity prices, and if the tariffs fall, they will use the electricity they produce themselves. Some of them see the installation as a sort of savings account, a payment in kind. The flow of money is transformed into a flow of electricity. The postponed use of the stored electricity, and more generally the possibility of using the electricity produced, turns the installation into what one might call a "giant piggy bank." Similar practices have been observed in relation to vegetable gardens (Wéber, 1998) or to food storage by the poorer sections of the population (Perrin-Heredia, 2011). They allow one to manage and thus to cope when times get hard.

The interviewees thus considered solar panels in terms of the local electricity market and imagined how the latter might evolve. The purpose of installing your own rooftop PV panels is therefore to ensure continued consumption in a country where access (economic, should there be power outages) to electricity and to its consumption cannot be taken for granted.

The producers who are worried about future pension levels have a similar way of thinking, though they tend to anticipate a fall in income rather than an increase in outgoings. They see the benefits of their installations in terms of a secure electricity supply in an uncertain electricity market – even more so in overseas departments which for the time being have special status, and while things do not bode well for their lives as future pensioners. Such calculations place the installation of solar panels in the register of household budget management. They do not mention any environmental issues, even when the latter are of concern to them. However, what is important here is the shift toward non-fossil fuels which, in their mind, will guarantee access to less expensive electricity. The two registers – environment and economics – are not incompatible. They support one another.

STATUS OF SOLAR ENERGY AND ELECTRICITY CONSUMPTION

Existing literature has examined on-grid systems and household consumption and provides a mixed picture of consumption practices (Keirstead, 2007). An analysis of our observations and interviews provides some of the keys with which to carry out an anthropological examination of the link between electricity production and consumption.

Firstly, some of our interviewees' concerns relating to "electricity budget" management place electricity production within a broader range of energy management. For example, Henri has changed his windows and has created openings so as to reduce his use of air-conditioning. He believes that "*it's not because you use a lot of electricity that you opt for solar power. You do it so that you use and manage electricity more efficiently.*" As we have seen above, not everyone associates the installation of solar panels with limiting their electricity consumption. On the contrary, once the installations have been paid for through resale, the profits may cover electricity bills, without having to worry about reducing electricity consumption. So "making savings" (on the electricity bill) does not always mean an attempt to reduce consumption. "*We're not going to buy something in order to cut back (…). There has to be more comfort, not less (…) and this is comfort, a real pleasure, and it's like a car on the island - cars are expensive, but if you don't have one, you're miserable, so there you go …*," answered one of our interviewees when asked if the panels were going to change her daily practices. For her, focused on her budget, PV panels allow the same level of electricity consumption without further expenditure; the future cannot mean restricted use of electricity, as this would be synonymous with less comfort and a form of social regression. This stance is understandable in relation to Reunion Island's economic situation, as set out above, that is, catching up in terms of the level of household equipment.

Secondly, to this contextual data must be added the collective psyche associated with solar energy. For Raineau (2011), renewable energies and fossil fuels relate to different energy systems (centralized/decentralized); the energy source thus leads to technical choices based on a certain relationship with the world and which result in a specific relationship with nature. Dobigny (2008) thinks along the same lines, observing that within a dwelling, ENRs modify the attitude and practices of the residents, who adapt to the new sources of energy (in relation to solar water heaters). In the present case, these representations would not appear to lead to restricted practices. Some producers consider that solar energy is free and available and do not distinguish solar panels from solar thermal heating: "*There's a source of energy that is available and free, the sun, so you install a solar-powered water heater for your swimming pool*" said Xavier. The interviewees thus envisage using solar energy for purposes they deem to be superfluous (e.g., heating a swimming pool). This is true of all producers, whatever the rationale behind their installation may be (environmental, budget management or access to electricity).

Thirdly, the status accorded to electricity varies, depending on whether it is for personal consumption or resale, that is, whether or not it has the status of merchandise. Kopytoff (1986) talks about the "cultural biography" of objects, suggesting that an object's status changes, depending on the stage of its circulation and on the sphere (commercial, gift, etc.). In this case, electricity drops its status as merchandise when it is privately consumed. In this respect, the discourses on batteries and on the electricity stored within are a powerful example. As far as the interviewees were concerned, if the electricity produced is not on the grid and is not being used by EDF, this almost physically removes it from the marketplace and thus from the monetary sphere: "*If someone tells me tomorrow that during the night I can use the electricity which is in the batteries, on the basis that it doesn't cost more because we know that it's not going to be used, I'll run the dishwasher and the washing machine, etc. at night ...*" (Philippe). The electricity is referred to as "free," with no consideration for the loss of earnings by not selling it on. These processes of commodifying and decommodifying electricity have a potential impact on the use to which it is put. In the present case, the notion that the electricity produced is "free" allows people to make more intensive use of certain appliances, or to use electricity in ways which until now have been excluded. Environmentally speaking, there is the risk of modifying norms of comfort and of increasing electricity consumption which will have to be borne by the grid (thus creating recourse to fossil fuels) when solar production is insufficient to meet these new needs. The use of solar electricity is therefore not free of certain environmental risks.

Lastly, some PV producers with environmental concerns say their level of electricity consumption is already so low that they cannot reduce it any further. They say this despite the fact that their overall energy consumption is high and that they have a boat, a 4-wheel drive car, or one or more motorbikes. Here we might mention the "social scenes" introduced by Wéber (2012) which aim to take account of the diversity of reference systems used in everyday economic calculations (of which energy consumption is one). These systems refer to distinct spheres, such as the domestic housing sphere, which exists alongside that of transport and leisure activities. Applied in this case, it means that energy consumption (including electricity) does not seem to be assessed in the same manner – and depending on the "social sphere" in question, may not be assessable at all. In this respect, some producers do not seem to see any incompatibility between a proactive environmental approach, which takes the concrete form of rooftop PV panels, and high levels of energy consumption (gas included).

CONCLUSION

The decision of whether or not to invest in solar panels mobilizes modes of "ordinary practices of calculation" (Wéber, 2010) which relate to the particular context of electricity in Reunion Island (fossil energy and, in terms of comfort, catching up with electricity markets in mainland France and in French Overseas Departments). The choice of PV panels made by households is based largely on a desire for independence (no more fossil fuels and electricity cuts) and access to electricity. So over the course of the analysis, arguments relating to the environment took a more financial/budget turn, linked to the producers' desire to maintain or increase the number of activities requiring electricity. Islanders thus re-appropriate technical devices in accordance with their specific priorities and constraints.

These calculations appear to varying extents in the islanders' discourse and help to define the purpose of the electricity production. Environmentally motivated producers thus claim to have made a "cash-neutral transaction"; they feel that making a financial profit from selling green electricity is not compatible with their commitment to the environment. They make no link between their electricity production and their electricity consumption. As far as they are concerned, their production supplies the island's electricity network with green energy. Whether or not they use it is immaterial.

Not all of the producers we met feel this way. Against all economic logic, some producers want to use all of their self-generated electricity (thus undermining their purely financial interests) and show no real interest in green energies. They place electricity production in the domestic domain. They thus position themselves first and foremost as consumers who do not wish to endure power outages, or who are trying to finance their electricity bills by becoming producers. The latter case is an example of calculation logic which prioritizes electricity consumption and not the "green nature" of the electricity produced. Processes of "earmarking" (Zélizer, 1994) the money from resale thus allow citizens to maintain their level of electricity usage for the same "electricity budget." Producers thus draw upon lay economic skills relating to budget management (monthly payments for the purchase of the PV panels, which do not burden the budget; paying "once and for all" to eliminate all future bills) rather than to maximizing profit.

The purpose of these skills is to maintain a level of comfort relatively new in Reunion Island. The local context is one of playing catch-up with mainland France. These producers therefore see any restriction to electricity use as a form of social regression. More generally, energy policies treat levels of comfort as an expression of individual choice – just like economists, cf. Dubuisson-Quellier (2010) – and focus their actions on the end-consumer (Wilhite, 2008); whereas these levels relate to social and cultural norms. Such policies thus place responsibility for climate change on the shoulders of individuals who are trapped within a consumer system. In an insular context such as that of Reunion Island, catching up comfort has led to the increased use of fossil fuels. The islanders are thus torn between norms which encourage ever-increasing consumption, so as to be part of a mass consumer society, and the need to consume less in order to comply with environmental policies. Their representations of the electricity produced ("clean" and "free"; commoditization/decommoditization) allow them to make conservation of the environment compatible with electricity uses that are part of social consumption norms. The "social scene" concept (Wéber, 2012) also makes it possible to clarify modes of adjustment between environmental values and everyday energy-hungry practices (for the entire spectrum of fuel types).

In all cases, the interviewees were very much energy-aware (its rarity, its origins, etc.). The producers' calculations engage a vision of the future where the environment will be more or less of an issue. While they have in common the fact that they are considering the electricity landscape of the future, they do not share identical visions: some foresee greater environmental problems (relating to fossil fuels), others see grid tensions (due to increased demand), or market changes (higher electricity prices, lower feed-in tariffs), or personal financial problems (such as retirement).

The purchase of PV panels thus enters the realm of anthropological research and feeds existing reflection on the diversity of economic rationalities. The concepts of the discipline ("ordinary practices of calculation," "money earmarking," "cultural biography," commoditization/decommoditization, "social scenes", etc.) allow us to analyze the cultural dimension of the modes of calculation at stake when installing solar panels, far removed from the neoclassical economic theories upon which public policies are based. Such insight is important for understanding how people respond to policy appeals for PV panels. Anthropology therefore has an important role to play in the debate on energy transition.

NOTES

1. *Source*: http://www.developpement-durable.gouv.fr/IMG/pdf/LPS36.pdf

2. There is a relative abundance of works on the cultural aspect of energy practices. These references are cited as examples. See Strauss, Rupp, and Love (2013) for a review of the literature on the topic.

3. *Source*: Institut National de la Statistique et des Etudes Economiques – Insee. http://www.insee.fr/fr/themes/document.asp?reg_id=24&ref_id=18713#e1

4. *Source*: Plan Régional des Energies Renouvelables et d'Utilisation Rationnelle de l'Energie – Prerure. http://www.regionreunion.com/fr/spip/IMG/pdf/prerure_presentation.pdf; http://www.regionreunion.com/fr/spip/spip.php?article1829

5. *Source*: ARER. http://www.arer.org

6. A total of 29% fall in obligatory purchase tariffs between 2011 and 2013 (0.46 €/kWh in March 2011 and 0.3266 €/kWh in July 2013); 20-year repurchase agreement.

7. Taxation: 2012 tax credit of 11%, 16,000 euro ceiling for a married couple with no children. Approximate maximum of 1,000 euros for a 3 kWc installation; the region offers an "energy cheque" of 3,000 euros.

8. About 750,000 American households according to Nick Rosen (*Off the Grid: Inside the Movement for Space, Less Government, and True Independence in Modern America*). Courrier International, 1036.

9. The decree dated April 17, 2009, pertaining to the thermal characteristics of residential buildings in French Overseas Departments, stipulates that in new buildings hot water must henceforth be produced by solar energy.

10. *Source*: Insee.

REFERENCES

Akrich, M. (1987). Comment décrire les objets techniques? *Techniques et Culture*, *1*(9), 49–64.

Chappells, H., & Shove, E. (2005). Debating the future of comfort: Environmental sustainability, energy consumption and the indoor environment. *Building Research Information*, *33*(1), 32–40.

Debourdeau, A. (2011). Domestiquer le solaire. In S. Barrey & E. Kessous (Eds.), *Consommer et protéger l'environnement* (pp. 47–66). Paris: L'Harmattan.

Dobigny, L. (2008). Changement énergétique et rapport au monde. In M.-J. Menozzi, F. Flipo, & D. Pécaud (Eds.), *Energie et Société. Sciences, gouvernances et usages* (pp. 215–224). Saint-Remy-de-Provence: Edisud.

Dubuisson-Quellier, S. (2010). La consommation comme pratique sociale. In P. Steiner & F. Vatin (Eds.), *Traité de sociologie économique* (pp. 725–776). Paris: P.U.F., Collection Quadridge.

Henning, A. (2000). *Ambiguous artefacts. Solar collectors in Swedish contexts. On processes of cultural modification* (p. 44). Stockholm: Stockholm Studies in Social Anthropology.

Henning, A. (2005). Climate change and energy use. The role for anthropological research. *Anthropology Today*, *21*(3), 8–12.

Henning, A. (2008). Heating Swedish houses: A discussion about decisions, change and stability. *Anthropological Notebooks*, *14*(3), 53–66.

Keirstead, J. (2007). Behavioural responses to photovoltaic system in the UK domestic sector. *Energy Policy*, *35*, 4128–4141.

Knight, D., & Bell, S. (2013). Pandora's box: Photovoltaic energy and economic crisis in Greece. *American Institute of Physics Journal of Renewable and Sustainable Energy*, *5*(3), 1–16.

Kopytoff, I. (1986). The cultural biography of things: Commodization as process. In A. Apparudai (Ed.), *The social life of things. Commodities in perspectives* (pp. 64–91). Cambridge: Cambridge University Press.

Lutzenhiser, L. (1992). A cultural model of household energy consumption. *Energy*, *17*(1), 47–60.

Perrin-Heredia, A. (2011). Faire les comptes: Normes comptables, normes sociales. *Genèses*, *3*(84), 69–92.

Pinçon, M. (1986). Autoproduction, sociabilité et identité dans une petite ville ouvrière. *Revue Française de Sociologie*, *XXVII*, 629–653.

Pruvost, G. (2013). L'alternative écologique. Vivre et travailler autrement. *Terrain*, *60*, 36–55.

Raineau, L. (2011). Vers une transition énergétique? *Natures Sciences Sociétés*, *19*(2), 133–143.

Savidan, L., Shaffar, A., Dimou, M., & Garde, F. (2008). La consommation énergétique des ménages à la Réunion: Vers un retour à l'autonomie par le biais des énergies renouvelables. *Etudes Caribéennes*, [on line], 11 | Décembre 2008, URL: http://etudescaribeennes.revues.org/3519

Shove, E. (2003). *Comfort, cleanliness and convenience*. Oxford, Berg: The Social Organization of Normality.

Shove, E., Chappells, H., Lutznhiser, L., & Hachett, B. (2008). Comfort in a low carbon society. *Building Research and Information*, *36*(4), 307–311.

Strauss, S., Rupp, S., & Love, T. (Eds.). (2013). *Cultures of energy. Power, practices, technologies*. Walnut Creek, CA: Left Coast Press.

Veblen, T. (1899). *The theory of the leisure class*. New York, NY: Penguin.

Wilhite, H. (2008). New thinking on the agentive relationship between end-use technologies and energy using practices. *Journal of Energy Efficiency*, *1*(2), 121–130.

Wilhite, H., Nakagami, H., Masuda, T., Yamaga, Y., & Haneda, H. (2001). A cross-cultural analysis of household energy-use behavior in Japan and Norway. In D. Miller (Ed.), *Consumption: Critical concepts in the social sciences* (Vol. 4, pp. 159–177). London: Routledge. (Reprint of an article originally published in *Energy Policy*).

Wilhite, H., Shove, E., & Lutzenhiser, L. (2000). Twenty years of energy demand management: We know much about individual behavior, but how much do we really know about demand. *ACEEE, summer study on energy efficiency in buildings*.

Winther, T. (2008). *The impact of electricity. Development, desires and dilemmas*. New York, NY: Berghan Books.

Winther, T., & Bouly de Lesdain, S. (2013). Electricity, uncertainty and the good life – A comparison of French and Norwegian households responses to policy appeals for sustainable energy. *Energy and Environment Research*, *3*(1), 71–83.

Wéber, F. (1998). *L'Honneur des jardiniers. Les potagers dans la France du XXe siècle*. Paris: Socio-histoire.

Wéber, F. (2009). *Le Travail à-côté. Une ethnographie des perceptions*. Paris: Éditions EHESS. Coll. "En temps & lieux".

Wéber, F. (2010). Le calcul économique ordinaire. In P. Steiner & F. Vatin (Eds.), *Traité de sociologie économique* (pp. 367–407). Paris: PUF collection Quadridge.

Wéber, F. (2012). Séparation des scènes sociales et pratiques du calcul. A la recherche des raisonnements indigènes. In N. Coquery, F. Menant, & F. Wéber (Eds.), *Ecrire, compter, mesurer. Vers une histoire des rationalités pratiques* (pp. 65–85). Paris: éditions de l'Ulm.

Zélizer, V. (1994). *The social meaning of money*. New York, NY: Basic Books.

"BUG WOOD": CLIMATE CHANGE, MOUNTAIN PINE BEETLES AND RISK IN THE SOUTHEASTERN BRITISH COLUMBIA LOGGING INDUSTRY

Patrick B. Patterson

ABSTRACT

Purpose – *The purpose of this paper is to analyze how the culture in the logging industry in the East Kootenay/Columbia region in British Columbia, Canada, is changing as warm winters resulting from climate change drive expansion of a native tree-killing pest, the mountain pine beetle* (Dendroctonus ponderosae).

Methodology/approach – *The paper is derived from historical records and 11 months of ethnographic fieldwork conducted from July 2010 to May 2011.*

Findings – *This analysis found that the insect outbreaks are generating a heightened sense of economic and physical vulnerability in the logging industry, undermining previous assumptions of sufficiency and confidence.*

Climate Change, Culture, and Economics: Anthropological Investigations
Research in Economic Anthropology, Volume 35, 47–64

ISSN: 0190-1281/doi:10.1108/S0190-128120150000035003

Research limitations/implications – *This paper presents results from a study of a specific region, and caution should be used when comparing these results with similar phenomena in other contexts.*

Social implications – *The forest industry is an important employer throughout the British Columbia interior; the cultural changes documented here indicate that climate change, manifested in insect outbreaks, is generating cultural dislocation that can have negative consequences beyond the immediate economic impacts.*

Originality/value – *This paper provides a detailed analysis of how an unanticipated consequence of climate change is driving adjustments in a subculture in a technologically advanced society.*

Keywords: Climate change; Mountain Pine Beetle; risk; forestry; culture; British Columbia

The Mountain Pine Beetle (*Dendroctonus ponderosae*) is a small, 5 mm long, nondescript black insect that is native to the forests of British Columbia, Canada (Gadd, 2009; Nikiforuk, 2011). They spend most of their time burrowing in the water and nutrient-rich inner bark of pine trees, where they lay eggs and are preyed upon by woodpeckers and other insectivorous birds (Gadd, 2009; Logan & Powell, 2001). Compared with more famous inhabitants of the British Columbia forest, such as grizzly bears, mountain caribou, and spotted owls, mountain pine beetles are easy to overlook and it is difficult to see how they could fundamentally change patterns of work and ideas around risk in the forest industry in an industrialized society. In this paper I argue that is exactly what the mountain pine beetles have done.

Mountain pine beetles kill the trees they inhabit, either directly by removing so much of the water-carrying layer that the trees are "girdled" and starve, or indirectly by introducing a virulent type of blue-stain fungus (*Grosmannia clavigera*), which is parasitic on the tree's nutrient-carrying system and ultimately starves the tree (Logan & Powell, 2001). Loggers in the British Columbia interior call beetle-killed stands of trees "bug wood" and since the late 1980s infestations have become increasingly common and of much greater extent than they had previously been (Carroll, Taylor, Régnière, & Safranyik, 2003; Taylor & Carroll, 2003). Very cold winter weather is the most important control on mountain pine beetle populations (Gadd, 2009; Logan & Powell, 2001) and scientists now attribute the expansion of these beetles in North America to warmer winters generated by global climate change (Carroll et al., 2003; Logan & Powell, 2001; Mitton & Ferrenberg, 2012; Stahl, Moore, & McKendry, 2006).

To say that the mountain pine beetle outbreaks in the British Columbia interior over the past 20 years caused severe damage to the forest that is available for commercial logging is an understatement. Lodgepole pine (*Pinus contorta*) is a very common native tree species, which is suitable for lumber and pulp production and makes up a large portion of the timber harvested in the British Columbia interior (BC Ministry of Forests, Mines, and Lands, 2010; Drushka, 1998, 1999). It is also sun-tolerant and fast growing and lodgepole pine seedlings were planted to reforest many logging sites as an investment toward future lumber production (BC Ministry of Forests, Mines, and Lands, 2010; Kimmins, 1997; Parish & Thompson, 1994). According to the British Columbia Ministry of Forests, Mines and Lands (or Ministry of Forests), in the two decades preceding 2008 mountain pine beetle infestations killed approximately 620 million cubic meters of lodgepole pine and they predicted that by 2015 70 percent of all mature lodgepole pine trees in the province would be killed (BC Ministry of Forests, Mines, and Lands, 2010). Although trees killed by mountain pine beetles can be harvested for lumber and pulp, bug wood is only usable in the first four to five years after the tree dies, before the trunk is damaged by rot. The people who work in the interior forest industry have had to adapt in several ways to the new realities posed by the mountain pine beetle outbreaks and their aftermath.

In an essay Clyde Kluckhohn commented that, "every culture is a precipitate of history. In more than one sense history is a sieve. Each culture embraces those aspects of the past which, usually in altered form and with altered meanings, live on in the present" (Kluckhohn, 1971, p, 28). The research presented here takes that longer context into account. The data are derived from historical accounts of the British Columbia forest industry, as well as from an extended period of ethnographic field research conducted in the East Kootenay/Columbia region in British Columbia, Canada, in 2010–2011. Although the impacts of warmer weather and the mountain pine beetle outbreaks were not the focus of the original study, during the field research manual fallers and other people employed in the logging industry often referred to problems posed by working in the standing dead trees left after an infestation. This paper analyzes impacts that the beetle outbreaks are having on meanings people in the East

Kootenay/Columbia logging industry place on the forest and on their work. I argue that these manual fallers, and people in the wider forest industry in British Columbia, provide a glimpse into ways that unanticipated consequences of climate change impact local cultural expectations and norms.

METHODS

The data that are analyzed and presented in this paper are derived from a project funded by WorkSafe BC to conduct a study of the safety culture among manual tree fallers in the British Columbia logging industry (Rock & Patterson, 2013). This case study (Keddie, 2006) involved an 11-month period of ethnographic field research (Hammersley & Atkinson, 1995; Liamputtong & Ezzy, 2005) from July 2010 to May 2011 in the East Kootenay/Columbia region. That area is in the Rocky Mountains, between the Purcell Mountains and the Kootenay Range, and is characterized by vertical terrain where manual tree felling methods are a necessary part of logging activities. Conducting the study there ensured that there would be a suitable population of potential study participants. Before commencing the field study, the research protocol was submitted to the Conjoint Faculties Research Ethics Board at the University of Calgary and received approval. The research process included obtaining informed consent from all participants and protecting their anonymity. All of the names given throughout this paper are pseudonyms.

Based on the literature describing the structure of the forest industry in the British Columbia interior (Drushka, 1999; Patterson, 2008; Resource Systems Management International, Randy Sunderman, & Associates, & Guidelines Consulting, 1999), participant recruitment was planned around purposive sampling (Oliver, 2006) and supplemented with recruitment during chance encounters while residing in the area. The goal of the purposive sampling was to ensure that the study included the perspectives of people who worked with manual fallers and could provide insights, as well as manual fallers themselves. By the conclusion of the field research the sample included seven manual fallers who were working at the time in the logging industry, two machine operators, five logging contractors, one administrator employed by a logging company, two managers employed by forest product companies, and one retired forest product company manager. One manual faller and one machine operator also provided follow-up interviews to discuss some of the preliminary findings from the study. Fourteen of the participants gave consent to be digitally recorded and six were recorded through field notes.

The recorded sessions were transcribed and all of the data were analyzed using a thematic approach (Bernard, 1995; Hammersley & Atkinson, 1995; Liamputtong & Ezzy, 2005). The results of that analysis are available through WorkSafeBC (Rock & Patterson, 2013). The consequences of the mountain pine beetle outbreaks in the East Kootenay/Columbia were recurring topics that were identified for additional attention. In the follow-up analysis, the previous, more general, analysis of the participants' orientations toward workplace safety was treated as the starting point for comparisons, along with the ethnographic and historical literature describing the culture in the British Columbia interior logging industry (Arruda, 1999; Cartwright, 2007; Drushka, 1998; Patterson, 2008). I then searched the data for examples where the study participants mentioned topics that are related to the phenomenon of mountain pine beetle infestations and salvage logging. I then examined those passages for common themes.

FROM SUSTAINED YIELD TO SALVAGE

The British Columbia forest industry is based on an approach to forestry that has its roots in 19th century Europe, where foresters applied scientific methods to the management of forests. Their model was to harvest trees, plant seedlings to replace them, and to control potential threats to wood production, such as fires, disease, and insect infestations (Drushka, 1999, 2003; Fisher, 1907). Shortly before World War I the government of British Columbia, influenced by the forest conservation ideas espoused by Bernard Fernow and other European-born foresters, opted to retain title over forested Crown Lands, rather than sell them (Drushka, 1999). The government's goal was to use the forest as a long-term source of resource revenues and as a way to generate employment (Drushka, 1999). That ownership structure is still in place and the British Columbia government holds the title to 95 percent of the forested land in the province (BC Ministry of Forests, Mines, and Lands, 2010). The provincial Ministry of Forests allows private companies to acquire timber from Crown land through a system of harvesting licenses (BC Ministry of Forests, Mines, and Lands, 2010; Drushka, 1999).

Under the harvesting license system the Ministry of Forests applied the concept of "sustained yield," wherein annual timber harvest amounts are based on the projected volume of new timber that the harvestable forested land base grows each year (Drushka, 1999). Following logging, cut areas are replanted with native tree species to ensure future wood supplies (BC Ministry of Forests, Mines, and Lands, 2010; Drushka, 1999; Robson, 1995). Forest product companies are required to submit plans to the local Ministry of Forests officials for their logging operations on Crown land and until 2003 they were required to harvest a set amount of their available timber volume each year or forfeit part of their future allocations (Queen's Printer, 2003). In turn, to reduce variability in logging services availability, quality, and cost forest product companies normally hire the same local logging contractors from year to year and negotiate contracts directly with them rather than using an open tender bidding system (Patterson, 2008; Rock & Patterson, 2013). Some British Columbia interior forest product companies formalized the on-going relationships with specific logging companies by providing "evergreen" contracts that assure the contractor a specified volume from the annual harvest allocation. The long-term local business links within the forest industry fostered by the sustained yield model also increased familiarity and informal communication between contractors and logging operations managers (Rock & Patterson, 2013). That sense of long-term continuity in the British Columbia interior logging industry was reinforced because the companies were often, literally, family concerns, as contractors employed their sons in their companies and eventually turned over ownership to them (Patterson, 2008).

In the British Columbia interior, including in the East Kootenay/Columbia region, long-term working relationships remain the norm, as logging contractors recruit workers from the local community and seek to keep experienced people in their crews (Patterson, 2008; Rock & Patterson, 2013). While transient forestry workers living in temporary camps are part of the folklore and history in western North America (Drushka, 1992, 1998; Felt, 1963; Grainger, 1996; Harrison, 1982), people in the British Columbia interior logging industry expect to live and work within a single local area for an extended period of time, often for their entire careers (Arruda, 1999; Drushka, 1998; Patterson, 2008). The experiences reported by the participants in this study followed that norm and, with two exceptions, all of them had lived in their communities and worked in the logging industry for more than 10 years. Over half of the participants reported that their fathers, and in several cases grandfathers, worked in the local logging industry.

In the logging industry subculture in the East Kootenay/Columbia region, most of the participants in this study reported that they spend recreational time in the forest, as well as working there. For example, Rod, a manual faller who grew up on a farm in the same community where he now resides, said, "Even if I'm not logging, I'm hunting or fishing ... You know, it's very seldom that I'm in the house. Or I'm at the cabin, fishing. It fits my lifestyle and my family's as well; they're all into it too." Other participants expressed a similar sense of appreciation and familiarity with the areas where they worked. Tom, a machine operator, said, "in my particular job, I get to know every little road and every little place that there is around here." He went on to explain:

> I'm quite interested in the environmental part of it, the birds and the flowers and the fossils and the, "why is this that way" and so on, and "where does that little trail go?" so I'm always poking around and learning something, figuring something ... the animals of course, we know most them and their habits just because I've spent all my life chasing them around and observing them, and what have you. The birds and various flowers and plants and this kind of stuff is interesting, and it all kind of ties into the big picture, so when I go out into the bush it's a, "what's new today?" kind of thing. There's always something. Even the same place changes, like now, the leaves are turning, it rains, the creek's up.

While the sense of place and appreciation of the forest the loggers express is based on personal experience and does not necessarily equate with scientific concepts of ecological protection and biodiversity maintenance (Kimmins, 1997), it does indicate that the region's loggers have a sense of identification with the forest.

This sense that the forest has value to loggers aside from its economic contribution to their livelihood presents an apparent paradox, since commercial logging involves cutting trees in large numbers and could be viewed as forest destruction (Peace, 1998; Satterfield, 2002; Stefanick, 2001). Loggers in British Columbia resolve the conflict for themselves by taking a pragmatic view of forest stewardship, which emphasizes the forest's capacity to grow new trees when logging is done using appropriate methods. The loggers base their sense of successful stewardship on their own experiences, where within their careers older participants have seen bare cut-blocks they logged, replanted, and covered with new trees that have reached 10 meters in height and 15 cm or more in trunk diameter. Those trees are not fully developed; even lodgepole pine, the fastest-growing coniferous tree native to British Columbia require 60 years to reach maturity, but they are seen by loggers as evidence that the system works. For example, Henry, a retired logger, noted that the second-growth forest in areas he logged in the 1960s is "coming along well" and that most people who see it do not even realize it was logged. Loggers in the East

Kootenay/Columbia region want to see forests successfully re-established following logging operations so that there will be continue to be a logging industry in the areas they view as their home.

In the British Columbia interior, mountain pine beetle outbreaks have turned the long-term resource represented by healthy, growing trees into one that is dead and has a four or five year window for economic value before the beetle-killed trees rot and fall down. At the institutional level of Ministry of Forests policy and forest product company harvesting plans, the logging industry in the British Columbia has changed in response to the mountain pine beetle's expansion. Foresters in British Columbia dealt with local insect outbreaks by quickly logging the affected area, or by conducting "prescribed burns" where fires were deliberately set to kill insects in areas with infested trees (Nikiforuk, 2011). Foresters also attempted to control mountain pine beetles in smaller infested areas by conducting "fall-and-burn" operations in the winter. The objective in a fall-and-burn operation is similar to a prescribed burn on a small scale, and manual fallers work in an area felling infested trees into piles and burning them while the beetle larvae remain dormant. In the past such techniques were sometimes effective in stopping localized insect outbreaks. In this study, Henry commented that there had not been much bug-killed forest during his working career, except for a spruce budworm (*Choristoneura freemani*) outbreak. He said that in that case they were able to get into the area and "log like hell" for two years to clear out the infestation and salvage the wood. In contrast, the current mountain pine beetle outbreaks are on a much larger scale. Several of the manual fallers said that they occasionally worked in fell-and-burn operations for the Ministry of Forests, but efforts in beetle-killed areas now focus on salvaging as much wood as possible before it becomes unusable for commercial purposes.

While salvage logging provides profitable short-term work for logging contractors and their employees, with operations continuing even during the severe recession from 2008 to 2010 (BC Ministry of Forests, Mines, and Lands, 2010), it is contrary to the deep orientation in their subculture to seek long-term reliability. For example, Dan, a logging contractor, said:

> Where we're at right now, a lot of times where I go logging, I'm offended about it [the clear-cut logging methods being used], too, because we're taking everything out of that valley. Like, right now where we're logging, we've nicknamed it 'the moon' because there's the odd larch tree left standing through acres and acres up there, and the stuff that was there that we went in to log was all bug-kill, real bad. We just managed to salvage it. A lot of it went into pulp because it was too far gone for saw logs and other than that it was just blown over. It probably would have ended up getting a fire into it.

> A lot of people don't understand why we're actually logging in that area, why we're clear-cutting that area, why we're taking so much of it. They don't understand those things. But that's what bothers me the most is, "why are you taking all of this jack pine?" We're taking it because the bugs are in it!

Dan's comment shows his discomfort at having to cut timber at rates that he knows are beyond what is sustainable, and his frustration at having his concern for the forest questioned by people outside the logging industry. Phil, a relatively young logging contractor who has an adult son working in his logging crew, said that he is considering selling his equipment and closing his business. Other participants also said that there is little point in investing in their businesses if no one will be willing to buy them out when they retire.

WORSE AND WORSE – WORKING SAFELY AMONG DAMAGED TREES

Using chain saws for manual tree felling has long been recognized as a dangerous occupation (Forest Safety Task Force, 2004; Mason, 1976; Patterson, 2008; Slappendel, Laird, Kawachi, Marshall, & Cryer, 1993; Sundstrom-Frisk, 1984; West, Shkrum, & Young, 1996), and managing hazards is part of the culture in the logging industry. Province-wide mandatory training requirements have only been in place since 2005, and only for high-risk occupations such as manual felling and logging truck driving (BC Forest Safety Council, 2008; Rock & Patterson, 2013). All of the participants in this study said that they learned their skills mainly through on-the-job experience. Over time, individual loggers develop knowledge about forest hazards, while at a wider level, the people doing the same task in similar conditions have similar experiences and reactions, producing the subculture around workplace risks (Rock & Patterson, 2013). The loggers in the study recognized connections between their actions, experience, and their knowledge of dangerous situations. For example, when he was asked how he learned to fell trees Rod said, "It was actually a lot of learning on the job, probably most of it trial and error … being practical and safe was my main objective." The loggers often described first-hand experience as the most reliable way to gain the knowledge needed for assessing the potential hazards posed by trees, weather and surrounding terrain. For example, when Chris, an experienced manual faller in his forties, was asked how he knew what was hazardous in his work he said:

> From having close calls. Or listening to other fallers ...[you] hear, "well this guy got whacked with a limb today, there's lots of loose limbs" or whatever, so then you're looking every time. And then when you do see one up there, if you have to wedge it or whatever, then you tap it a couple times and you're always watching, and then when that limb gets jarred loose, and you're not watching and it lands right beside you, like "Hell!" It makes you think ten times harder about what you're about. The next time you go to a tree you're going to watch that so that you don't get whacked.

Fallers also often identified less tangible, psychological qualities helping them work safely. Later in the conversation with Chris, he described how manual fallers approach working in the forest:

> They're cautious when they go to work each day. How I look at it, if I can't see something that I haven't seen [before] at least once or twice a week, then it's time for me to quit. Once I start figuring I know everything then it's time not to do it anymore; if there's something a little different with that tree, or whatever. You shouldn't, I don't think you should be out there doing it if you figure you're going to walk up to every tree and successfully walk away safe, you better have that in the back of your mind there that potential of something falling out of that tree, when you're making your undercut and the way it falls over. So you want to be aware of all that. That's how I am anyway. I can do everything by the book, but there's, it doesn't matter, there's always going to be something there that's going to jump out and surprise you.

Fallers identified being alert and focused on the task, having confidence to make judgements about how to approach unknown situations, and being willing to adapt to circumstances as qualities that were important. Finally, they described the importance of developing a sense of timing in work contexts, such as timing when to end a chain saw cut so that they leave enough wood to act as a hinge that will guide the tree's fall, and also the importance of timing in planning tasks so that the crew works together effectively and safely.

Salvaging dead trees introduces a sense of urgency for loggers because beetle-killed trees quickly become commercially worthless. Forest product companies and the logging contractors they employ have responded by using the logging methods that are the most time and cost efficient. Large tracked machines known as "feller-bunchers" are able to cut and stack trees very cost-effectively and "mechanized logging" is a common approach to removing beetle-killed forest. Logging machines can operate on flat ground and slopes up to a maximum grade of 55 percent, or 55 meters of height gain per 100 meters of horizontal distance (BC Ministry of Forests, 2012). Mountain pine beetles do not only infest trees growing in easily-accessed locations, however, and some beetle-killed areas are on steep slopes that require manual felling. When Gary, a logging contractor, was asked why a crew might be moved on to a site early, and put increased pressure on the manual fallers, he said, "it's just if they wanted to get some beetle out, that wasn't laid out [*planned ahead*] then they might have to move something in to get it." For loggers, and especially manual fallers, the time pressure in short-notice reactions to mountain pine beetle infestations is associated with mistakes and accidents.

Loggers also associate certain characteristics of trees with danger. In particular, as older or weakened trees in a forest die they often remain standing (Kimmins, 1997), and become what British Columbia loggers refer to as "snags." Working around snags has always been an aspect of logging operations and loggers recognize that they present additional hazards, compared with healthy trees, as the trunks, branches, and roots decompose, weakening their structures. For machine operators there is relatively little additional danger in working with rotten trees because the workers are protected by steel-reinforced machine cabs. If a tree breaks and part of it falls on the machine it is unlikely to injure the operator. In contrast, any object falling from a tree can cause serious injuries to a manual faller and they face considerable hazards from snags. When Steve, one of the fallers in this study, was asked about dangerous situations in his work, he said:

> Big rotten trees always give you a thrill because you can plan for the best scenario possible, but with mother-nature and the wilderness, the ground conditions and the weather at the time and the stability of the tree visually, compared to what it could be that you can't see, changes so much. I mean, you could make one cut into that tree and it can completely change. Like, I've felled trees where you just touch it to put the undercut in and they fall over. They look safe but it is actually just a shell that's leaning slightly. You have no control over it.

Over time, as they work with snags manual fallers develop knowledge about the external signs that a tree is dead and the hazards that a snag is likely to contain, as well as a tacit assumption that those trees are never completely controllable. As Steve's description indicates, fallers also have an emotional reaction to the uncertainty involved in working with snags.

While snags are distributed in small numbers throughout multi-age stands within a forest, lodgepole pines often colonize burned areas and other forest openings, and as a result form stands that tend to have relatively few other tree species and are often of relatively uniform age (Parish & Thompson, 1994). Mountain pine beetle infestations kill close to 100 percent of the trees in those areas, which become snags that the logging crew has to work with. Jim, an older faller, explained:

> Some timber types in the East Kootenays here you have jack pine, lodgepole pine. Talk about insect damage, pine beetle has been in there, and the lodgepole pine, after it has

> been dead for a certain number of years, about five years, the root system is completely deteriorated. That is very uncomfortable, those blocks; you are in there falling and you get some wind coming up, look out! There are many trees like that and maybe you take out the big ones and you remove them, but you get amongst the little guys and a five inch tree, five inch diameter, can do serious damage [*if it hits someone*]... They have had some accidents in the last nine years, and like we say as fallers, "no worries about the tree you're working on; it's the tree you didn't see that is going to get you."... You're in control of the tree you're cutting, but the ones behind you, those are the ones that can get you.

Each year after the trees die there is incrementally more damage from decomposition and weather, especially wind, increases the variability in mountain pine beetle-killed stands, making those areas more dangerous to work in. When Chris was asked what has led to the dangerous situations he has experienced during his career, he said, "Now it seems to be getting worse and worse. Like, with a bit more of the pine stands, because they're trying to attack them more and more because of the pine beetle. So then you're starting to deal with way more broken tops and that, that are hung up there, that do evolve and eventually become hazards." For Chris and other manual fallers like him in the British Columbia interior, working in the beetle-killed forest is increasing their sense of vulnerability. As Jim suggested, they cannot look in every direction at once.

DISCUSSION

Historical accounts of the British Columbia interior logging industry since the end of World War Two (Arruda, 1999; Drushka, 1998), and the ways that loggers in the East Kootenay/Columbia described their work during this field study reflect a subculture that emerged in the context of the government's sustained yield policy and a forest environment that appeared to be able to support it. In addition to managing the available timber volume, the system of government forest ownership and harvesting licenses shaped economic expectations. At the local level, forest product companies repeatedly hired reliable contractors to carry out their logging operations, while logging contractors came to expect to work with specific forest product companies for long periods of time and built the capacities of their companies to meet the main client's requirements. Although the British Columbia forest industry experienced cyclical fluctuations in demand for wood products (Drushka, 1999), the sustained yield system fostered a culture in the logging industry marked by a strong preference for reliable people and work methods (Patterson, 2008) and whose members had a deeply held assumption that the forest would provide long-term security.

Another outcome of the relative security from the sustained yield system is that people in the British Columbia interior logging industry normally live and work in the same place for a long time, often for decades, and come to hold positive a sense of attachment to the local forests, a sense of "home," beyond its economic contribution to their livelihood. Although the realities of forest use in British Columbia often diverge from scientific ecological conservation models, Tom's interest in "the big picture" and Henry's comment that trees in areas he logged are "coming along," and similar views expressed by other participants, indicate that pragmatic notions of forest stewardship are an established part of the logging industry subculture. For the loggers who participated in this study, and make their livings cutting trees and transporting logs, having their "tracks covered" by new tree growth on logging sites is seen as an indication that they have done their jobs well (see also, Robson, 1995).

Hazards faced during day-to-day work have also shaped the subculture in the British Columbia logging industry. Province-wide formal training programs for production workers are a new feature in the logging industry (BC Forest Safety Council, 2008; Rock & Patterson, 2013) and most loggers in the British Columbia interior learned from experience how to work effectively with the hazards posed by the forest environment. In the case of manual fallers in this study, they reported various ways that trees, weather and the surrounding terrain interact with the actions they take – some outcomes are the ones they anticipate, but others take the individual by surprise. The emphasis loggers place on learning from experience and recognizing that forests are difficult to predict has generated a subculture that emphasizes perpetual alertness and caution. Over several decades the ecological environment and industry practices remained stable enough that people in the logging industry felt that they could rely on continuity and that they knew how to handle challenges that emerged.

Climate change, manifested in the expansion of mountain pine beetle infestations in the East Kootenay/Columbia region (BC Ministry of Forests, Mines, and Lands, 2010; Carroll et al., 2003; Logan & Powell, 2001; Mitton & Ferrenberg, 2012; Taylor & Carroll, 2003), introduced new factors that challenge the continuity loggers expect. Although mountain pine beetles are native to the area, the extent that the outbreaks would reach was not initially apparent. People in the forest industry had previous experience managing infestations by tree-killing insects, and over the course of the mountain pine beetle outbreak they applied those methods, such as the "fell-and-burn" and clear-cut techniques. The methods were not effective, however, because cold winter weather no longer acted as the most effective, but until recently largely unrecognized, brake on mountain pine beetle populations (Carroll et al., 2003; Mitton & Ferrenberg, 2012; Stahl et al., 2006). The result was the rapid collapse of an important part of the British Columbia interior timber resource, despite the efforts put into scientific forest management and reforestation.

The responses of study participants to the changes brought by the mountain pine beetle outbreak show that the experience has been shaking some of their deeply held assumptions about their own capacities to manage in the world they inhabit. In their experience, and in the pragmatic view of forest stewardship in their subculture, loggers expect that trees will grow back on logging sites; they have witnessed forest regeneration in areas where they have worked and they trust that it will occur eventually. The question is: will it be soon enough? Dan and the other participants in the study recognize that mountain pine beetle outbreaks are killing the timber supply at rates well beyond what they see as sustainable; regardless of whether loggers salvage the dead trees, or the trees stay in the forest and eventually fall over and rot, the forest is less likely to be able to support their businesses and jobs in the longer term. The Ministry of Forests estimates that for several decades the annual timber harvest in British Columbia will have to be reduced by approximately 30 percent to 50–60 million cubic meters, from the 77 million cubic meter average that sustains the forest industry at its current level (BC Ministry of Forests, Mines, and Lands, 2010, p. 135). Although the British Columbia forest industry has faced a series of challenges since its peak in the mid-1980s (Barnes & Hayter, 1997; Cashore, Hoberg, Howlett, Rayner, & Wilson, 2001; Drushka, 1999; Hayter & Barnes, 1997) and mountain pine beetle outbreaks are not the only sources of problems facing logging contractors, the shift toward salvage harvesting and experiencing an unknown future profoundly threatens the sense of continuity in the logging industry subculture and contributes to a growing sense of nervousness. In the longer term, the logging industry is also being decentered, as well established contractors and workers consider abandoning it and leave, and fewer people in the following generation believe it is a good investment to enter it.

For those loggers in the East Kootenay/Columbia region who continue to work in the industry, it will also be in an increasingly unpredictable physical environment, as mountain pine beetle outbreaks kill large stands of trees and change working conditions in the forest. Salvage logging in areas infested by mountain pine beetles impacts several of the factors shaping how loggers view workplace hazards, especially for manual fallers who work directly with standing dead trees. Manual fallers face a double challenge in their daily work with bug wood. As the more time-efficient, and relatively safe, feller-buncher machines remove beetle-infested stands on the "easy" flat or moderately sloped areas, manual fallers have to maintain high paced production in steep ground, which can compromise their safety. As well, fallers increasingly have to work in extensive stands of dead trees that have unknown and potentially highly variable structural integrity. Manual fallers and other loggers learn from experience to be aware of hazards in the forest, and to take measures to manage them; informally managing risk is part of the established sub-cultural values. The recent extensive mountain pine beetle infestations have created a forest environment that is *all* relatively dangerous, which is eliciting a deep sense of unease in the logging industry subculture.

CONCLUSION AND IMPLICATIONS

It is important to keep in mind with case studies, including the one presented here, that they are limited in scope and insights are only likely to apply in similar contexts (Keddie, 2006). In this case, mountain pine beetle outbreaks extend across British Columbia, the regulatory structures governing the logging industry are uniform, and the patterns of contracting are widespread across the province (Drushka, 1999; Hayter & Barnes, 1997; Resource Systems Management International et al., 1999). This suggests that the insights are applicable to many other areas in British Columbia. Indeed, in areas where the mountain pine beetle infestations are even more extreme, such as near Prince George in the central interior, it is reasonable to assume that the impacts on the logging industry subculture are even more severe. Mountain pine beetles and closely related species are also native west of the Rocky Mountains in the United States of America (Logan & Powell, 2001; Nikiforuk, 2011), and people in the logging industries in those areas are likely to face similar ecological challenges, albeit in a context with different regulatory frameworks and business structures. It is also possible that as climate change makes northern regions more hospitable for mountain pine beetles they may expand into the Boreal forest east of the Rocky Mountains (Mitton & Ferrenberg, 2012). If that occurs, the effects on forests in those areas will be considerable, with a corresponding impact on human populations living there.

The impacts that mountain pine beetle expansion has had in the East Kootenay/Columbia region and British Columbia also provides a cautionary example showing how climate change has the potential to disrupt any industry that depends on consistent ecological conditions. Agriculture, fishing, pastoralism, and hunting are all obvious examples of livelihoods that may become increasingly vulnerable as local conditions change as part of global processes. What is just as troubling is that although urban populations no longer rely on local resources for their direct livelihoods, our current, technologically advanced societies remain based on having a relatively consistent environment supplying food and other basic necessities to urban markets. The interdependence in a globalized economy implies that on a more general level, climate change is likely to generate economic shocks and contribute to a growing sense of nervousness and unease in urban subcultures.

ACKNOWLEDGMENTS

This research is supported with funds from WorkSafeBC through the *Focus on Tomorrow* program. Dr. Melanie Rock held the WorkSafe BC Research at Work Operating Grant (RS2010-OG18) "'What was he thinking?': Cultural Factors in Manual Tree Felling Safety."

REFERENCES

Arruda, T. F. (1999). 'You would have had your pick': Youth, gender and jobs in Williams lake, British Columbia, 1945–1975. In *Beyond city limits: Rural history in British Columbia* (pp. 225–233). Vancouver: UBC Press.

Barnes, T., & Hayter, R. (1997). *Troubles in the rainforest: British Columbia's forest economy in transition, Canadian western geographical series* (Vols. 33). Victoria: Western Geographical Press.

BC Forest Safety Council. (2008). *BC forest safety council: Backgrounder*. Retrieved from http://www.bcforestsafe.org/files/files/about/History%20&%20Mandate%20of%20Council%20-%20Sept%2008.pdf; http://www.bcforestsafe.org/files/files/about/History%20&%20Mandate%20of%20Council%20-%20Sept%2008.pdf

BC Ministry of Forests. (2012). *Harvesting systems and equipment in British Columbia*. Victoria: British Columbia Ministry of Forests, Forest Practices Branch. Retrieved from http://www.for.gov.bc.ca/hfd/pubs/docs/sil/Sil468–1cb.pdf

BC Ministry of Forests, Mines and Lands. (2010). *The state of British Columbia's forests* (3rd ed.). Victoria: Forest Practices and Investment Branch.

Bernard, H. R. (1995). *Research methods in Anthropology: Qualitative and quantitative approaches* (2nd ed.). Walnut Creek, CA: Altamira Press.

Carroll, A., Taylor, S., Régnière, J., & Safranyik, L. (2003). Effects of climate change on range expansion by the mountain pine beetle in British Columbia. In T. L. Shore, J. E. Brooks, & J. E. Stone (Eds.), *Mountain pine beetle symposium: Challenges and solutions* (pp. 223–232). Victoria: Pacific Forestry Centre.

Cartwright, C. (2007). *Empty on the swan: A trucker and loggers view of the whiteswan road.* Cranbrook: Rocky Mountain Printers.

Cashore, B., Hoberg, G., Howlett, M., Rayner, J., & Wilson, J. (2001). *In search of sustainability: British Columbia forest policy in the 1990s.* Vancouver: UBC Press.

Drushka, K. (1992). *Working in the woods: A history of logging on the west coast.* Madiera Park: Harbour Publishing.

Drushka, K. (1998). *Tie hackers to timber harvesters: The history of logging in British Columbia's interior.* Madiera Park: Harbour Publishing.

Drushka, K. (1999). *In the bight: The BC forest industry today.* Madiera Park: Harbour Publishing.

Drushka, K. (2003). *Canada's forests: A history.* Montreal: McGill-Queen's University Press.

Felt, M. E. (1963). *Gyppo logger.* Seattle, WA: University of Washington Press.

Fisher, W. R. (1907). *Schlich's manual of forestry, forest protection* (2nd ed., Vol. 4). London: Bradbury, Agnew & Company.

Forest Safety Task Force. (2004). *A report and action plan to eliminate deaths and serious injuries in British Columbia's forests: A final report of the forest safety task force.* Vancouver: Workers' Compensation Board of BC.

Gadd, B. (2009). *Handbook of the Canadian Rockies* (2nd ed.). Jasper: Corax Press.

Grainger, M. A. (1996). *Woodsmen of the west.* Toronto: McClelland and Stewart.

Hammersley, M., & Atkinson, P. (1995). *Ethnography: Principles in practice* (2nd ed.). London: Routledge.

Harrison, P. H. (1982). Life in a logging camp. *BC Studies, 84*, 88–104.

Hayter, R., & Barnes, T. (1997). The restructuring of British Columbia's forest sector: Flexibility perspectives. *BC Studies, 133*, 7–34.

Keddie, V. (2006). Case study method. In V. Jupp (Ed.), *The Sage dictionary of social research methods* (pp. 20–21). London: Sage.

Kimmins, H. (1997). *Balancing Act: Environmental issues in forestry* (2nd ed.). Vancouver: UBC Press.

Kluckhohn, C. (1971). *Mirror for man: Anthropology and modern life.* New York, NY: McGraw-Hill Book Company.

Liamputtong, P., & Ezzy, D. (2005). *Qualitative research methods* (2nd ed.). Oxford: Oxford University Press.

Logan, J. A., & Powell, J. A. (2001). Ghost forests, global warming, and the mountain pine beetle (Coleoptera: Scolytidae). *American Entomologist, 47*, 160–173.

Mason, K. (1976). The effect of piecework on accident rates in the logging industry (incorporating a different approach to the exposure problem). *Journal of Occupational Accidents, 1*, 281–294.

Mitton, J. B., & Ferrenberg, S. M. (2012). Mountain pine beetle develops an unprecedented summer generation in response to climate warming. *The American Naturalist, 179*, E163–E171.

Nikiforuk, A. (2011). *Empire of the beetle: How human folly and a tiny bug are killing North America's great forests.* Vancouver: Greystone Books/David Suzuki Foundation.

Oliver, P. (2006). Purposive sampling. In V. Jupp (Ed.), *The SAGE dictionary of social research methods* (pp. 244–245). London: Sage.

Parish, R., & Thompson, S. (1994). *Tree book: Learning to recognize trees of British Columbia.* Victoria: Canadian Forest Service.

Patterson, P. B. (2008). Attributions to danger and responses to risk among logging contractors in British Columbia's southern interior: Implications for accident prevention in the forest industry. In D. C. Wood (Ed.), *The economics of health and wellness: Anthropological perspectives* (pp. 103–125). Amsterdam: Elsevier.

Peace, A. (1998). Anatomy of a blockade. Towards an ethnography of environmental dispute (Part. 2), rural new South Wales 1996. *Australian Journal of Anthropology, 10*, 144–162.

Queen's Printer. (2003, May 29). Forest (Revitalization) Amendment Act, 2003. Bill 29 – 2003. British Columbia.

Resource Systems Management International, Randy Sunderman and Associates, & Guidelines Consulting. (1999). *The contract harvesting industry in British Columbia: Final report.* Victoria: Science Council of BC and Forest Renewal BC.

Robson, P. A. (1995). *The working forest of British Columbia.* Madiera Park: Harbour Publishing.

Rock, M., & Patterson, P. (2013). *"What was he thinking?" Cultural factors in manual tree felling safety* (Rep. No. WorkSafeBC Research Grant RS2010-OG18). Richmond: WorkSafeBC Research Services.

Satterfield, T. (2002). *Anatomy of a conflict: Identity, knowledge and emotion in old-growth forests.* Vancouver: UBC Press.

Slappendel, C., Laird, I., Kawachi, I., Marshall, S., & Cryer, C. (1993). Factors affecting work-related injury among forestry workers: A review. *Journal of Safety Research, 24*, 19–32.

Stahl, K., Moore, R. D., & McKendry, I. G. (2006). Climatology of winter cold spells in relation to mountain pine beetle mortality in British Columbia, Canada. *Climate Research, 32*, 13–23.

Stefanick, L. (2001). Baby stumpy and the war in the woods: Competing frames of British Columbia forests. *BC Studies, 130*, 41–68.

Sundstrom-Frisk, C. (1984). Behavioural control through piece-rate wages. *Journal of Occupational Accidents, 6*, 49–59.

Taylor, S., & Carroll, A. (2003). Disturbance, forest age, and mountain pine beetle outbreak dynammics in BC: A historical perspective. In T. L. Shore, J. E. Brooks, & J. E. Stone (Eds.), *Mountain pine beetle symposium: Challenges and solutions* (pp. 41–51). Victoria: Pacific Forestry Centre.

West, R., Shkrum, M. J., & Young, J. G. (1996). Commercial logging fatalities in Ontario, 1986–1991. *American Journal of Forensic Medicine and Pathology, 17*, 299–304.

PART II
FOOD PRODUCTION/ PROCUREMENT

CONSERVATION UNITS, ENVIRONMENTAL SERVICES AND FRONTIER PEASANTS IN THE CENTRAL AMAZON: MULTI-FUNCTIONALITY, JUXTAPOSITION OR CONFLICT?

Ana Maria de Souza Mello Bicalho and
Scott William Hoefle

ABSTRACT

Purpose – *To critically assess the contribution of community-based logging, low-carbon emission non-timber activities, and direct payment for environmental services in building sustainable rural livelihoods in the Amazon.*

Methodology/approach – *Fieldwork undertaken in 2008, 2010, and 2013 on sources of income for 110 interviewed families living in and around three different types of conservation units located on the advancing frontier in western Pará State.*

Climate Change, Culture, and Economics: Anthropological Investigations
Research in Economic Anthropology, Volume 35, 67–105

ISSN: 0190-1281/doi:10.1108/S0190-128120150000035004

Findings – *Three scenarios identified with very different socio-environmental outcomes, (1) the multi-functional combination of agricultural and non-agricultural activities replaces frontier farming, reduces deforestation and carbon emissions, increases income, and promotes social inclusion, (2) the mere juxtaposition of green alternative activities alongside unsustainable frontier farming has limited regional impact, and (3) environmental restrictions interfere with rural livelihoods to the point that people to leave the countryside.*

Social implications – *Evaluation of the effectiveness of GO and NGO policies in one of the poorest and environmentally problematic regions of Brazil.*

Originality/value – *First-hand information at the family farmer level concerning sources of income from conventional and green land use systems which is essential for formulating viable socio-environmental policy capable of reducing deforestation and carbon emissions which negatively impact global climate change.*

Keywords: Multi-functional livelihoods; conservation and environmental services; Central Amazon; Brazil

INTRODUCTION

The Amazon region is considered to be a crucial environmental hotspot for preventing global climatic change. Ever since the planned colonization projects and massive road building programs of the 1970s, deforestation in northern Brazil has been condemned because it causes loss of biodiversity, interferes with precipitation patterns and river discharge, releases carbon dioxide through forest burning, and reduces the capacity for absorbing carbon (Davidson et al., 2012).[1] Therefore, environmentalists often consider frontier smallholders, ranchers, and commodity producers to be enemies of the planet and of mankind (Butler, 2010; GCP, 2007). Classic studies of this process also showed that colonization had a high human cost in the form of poverty, violent land dispute and accentuated social inequality (Foweraker, 1981; Hecht & Cockburn, 1990; Ianni, 1979; Oliveira, 1999; Schmink & Wood, 1992).

However, with the end of military rule in 1985 and the promulgation of a new constitution in 1988 many regional socio-environmental trends have been reversed.[2] Since this sea change in Brazilian politics, the land registry office INCRA (Instituto Nacional de Colonização e Reforma Agrária) has tried to defuse conflict on the advancing frontier by actively registering land,[3] the environmental protection agency IBAMA (Instituto Brasileiro do Meio Ambiente e dos Recursos Naturais Renováveis) has used remote sensing to more effectively enforce deforestation laws in real time and the administrator of conservation units ICMBio (Instituto Chico Mendes de Conservação da Biodiversidade) has shifted from biocentric preservationism which had no place for humans in conservation units to sustainable land use governance embracing historic frontier peasants in National Forests and Rubber Tapping Reserves set up with the express purpose of preventing the spread of conventional frontier slash-and-burn farming. This paper explores to what degree changed policy has actually reduced deforestation and benefited local people resident in or nearby three different types of conservation units located in western Pará, at the most westerly point of the advancing frontier in the eastern Amazon.

A MULTI-FUNCTIONAL APPROACH TO RURAL LIVELIHOODS IN THE BRAZILIAN AMAZON

A multi-functionality approach is employed to systematically compare conventional frontier land use to a combination of community-based forestry, crafts, tourism, and environmental service payments for residents within and outside of conservation units in the Central Amazon. To some anthropologists this may seem to be a curious interpretative strategy because multi-functionality is usually associated with productivist and post-productivist rural landscapes of Western Europe where rural geographers and sociologists developed this line of research to understand processes in which farmers, pressured by global markets and changing EU common agricultural policy reducing subsidies and encouraging agro-ecology, have adapted by taking on new agricultural and non-agricultural activities which together can sustain rural livelihoods. Multi-functionality eventually became official EU policy for depressed rural regions that had become redundant as primary-sector productive space where new consumptive activities like environmental services and tourism could become important full-time or part-time activities for farmers who needed new sources of income (Muirhead & Almas, 2012; Rennington, Renwick, & Burton, 2012; Wilson, 2010; Woods, 2011).

At the same time, the cultural and environmental turn in tourism created greater demand for this kind of post-productive consumption of the countryside whereby a growing number of urban people want to experience rural life and undertake recreation activities in green areas and conservation units. *If* compatible with farm hours and seasonality, rural people can take advantage of this new attitude to rent rooms and organize activities for city people who want to participate in what they think to be a low-stress daily routine on the farm (Cawley, 2010; Torres & Momsen, 2011). All of this may seem to be light years away from the reality of frontier peasants of the Amazon but ecotourism has been billed as a way of creating value for rural people living in or near conservation units. Local people possess first-hand knowledge of natural landscapes and native flora and fauna, and working as guides they can receive a form of payment for environmental services in lieu of previous unsustainable farming and hunting practices (Durham, 2008; Moisey, 2000; Stronza, 2008). This in turn dovetails neatly with 21st century conservation policy which has shifted from merely removing long-resident people from new conservation units and fining farmers who practice unsustainable activities in buffer areas to that of creating partnerships for devising community-based sustainable timber and non-timber forestry activities, with a special place for ecotourism, in order to achieve low-carbon growth, forest conservation, and alleviation of rural poverty (Eagles & McCool, 2000; Larson, Barry, & Dahal, 2010; Molnar, 2011; Taylor & Cheng, 2012).

These issues all come together in the Central Amazon where, since the 1990s as part of a national strategy to reduce deforestation and emissions, the Brazilian Ministry of the Environment has aggressively set aside enormous areas of the Amazon in different types of conservation units so that today about 25% of the region is encompassed by conservation areas and the goal for 2020 is 30% (MMA, 2013a). As almost all of the conservation units in the Amazon were established after 1970 many include a long-standing population composed of Amerindians and historic peasants with significant Amerindian ancestry. Full preservation units, such as National Parks and Ecological Stations, do not permit economic use or resident population, but National Forests, Rubber Tapping Reserves, and Amerindian Territories permit prior residents to stay if they use natural resources in a sustainable way. However, in practice this usually means that hunting, fishing, and farming activities are curtailed to subsistence levels and the combination of non-timber forestry, ecotourism, and payment for environmental services has been proposed as an alternative means of generating income for the population resident within conversation units and buffer areas (IBAMA, 2005a; MICT/MMA, 1994; MMA, 2002, 2013b).

So if conservation units are to be preserved and not suffer invasion by landless peasants it is utmost important to answer the question whether new full-time or multi-functional part-time rural activities can significantly contribute to Amazonian livelihoods. At the most basic level we can ask what kind of employment is created for local people? Can more income be earned in environmental services and in community-based sustainable forestry than in typical frontier farming? Is the work full-time and year round or merely part-time and seasonal? If the latter is the case, does the work conflict with the farm calendar or take up the slack in off-peak times of the year? Are these opportunities created on the expanding settler frontier where most deforestation occurs in the Amazon? In preserved parts of the Amazon located beyond the expanding frontier do the activities compensate rural activities curtailed by environmental policy inside and outside conservation units? Are institutional structures developed which permit the replication of successful experiments on a regional scale? Are all the social actors contemplated or do cultural fault lines appear which exclude certain groups? In sum, will the new activities save the Amazon rain forest and so contribute to diminishing global climatic change?

CONSERVATION POLICY FOR CONTAINING DEFORESTATION IN THE CENTRAL AMAZON

To answer these questions fieldwork was undertaken in 2008, 2010, and 2013 by a female rural geographer and a male cultural and social anthropologist in and around three different kinds of conservation unit in the Amazon. The three represent the range of conservation units present in the region: (1) the Amazon National Park (a full preservation unit which does not permit inhabitants that was invaded by settlers from other regions), (2) the Tapajós National Forest (a forest reserve whose administrators first tried to remove the historic peasants present but later shifted to devising sustainable activities for them), and (3) the Tapajós-Arapiuns Rubber Tapping Reserve (a more recent sustainable development unit that permits more economic activity which was set up at the request of the historic peasants).

The aim of the research was to critically evaluate suggested green solutions to socio-environmental problems of the Amazon using spatially nuanced micro-accounting of the actual contribution of different sources of income to livelihood strategies at the family level. The researchers did detailed interviews with 89 families resident in 26 communities spread across the conservation units and in the buffer zones concerning their use of land and resources, farming systems, labor regimes, market articulation, sources of monetary income, family structure, out-migration, access to public services, and political mobilization. For comparative purposes, similar interviews were undertaken with another 21 farmer families located far enough away from the conservation units to be uninfluenced by them (Table 1). The researchers also talked with park administrators, land office officials, and community leaders concerning environmental policy and the political resistance of the local population to avoid removal from the conservation units.

The three conservation units studied are part of a larger system located in a corridor of conservation units created in the late 1990s along the Cuiabá-Santarém Highway (BR-163) and the Transamazonian Highway (BR-230) with the express purpose of limiting deforestation in the Central Amazon (Fig. 1). The frontier has arrived in this part of the Amazon with increased logging, farming, and mining pressuring local peasants and Amerindians. National and international environmental concern over the consequent deforestation and carbon emissions, in turn, has given rise to greater regulation of conservation units by the Ministry of the Environment. In an attempt to prevent the construction of feeder routes off the main highways, which would open up new forest areas for settlement, a corridor of public forests occupies a 100-km deep strip alongside both sides of the BR-163. This involves 27 million hectares of forest, of which 1.52 million hectares have been allocated to sustainable forestry concessions for communities and private logging firms (MMA, 2008).

Table 1. Farmers and Workers Interviewed.

Farmer/Worker Type	Within Conservation Units	River Buffer Area	Road Buffer Area	Distant Road Non-Buffer	Road Near Santarém and Itaituba Non-Buffer
Historic peasant	31	14	0	0	0
Peasant settler	0	0	38	6	13
Rancher settler	0	0	1	3	0
Worker settler	0	1	3	0	0
Total	31	15	42	9	13

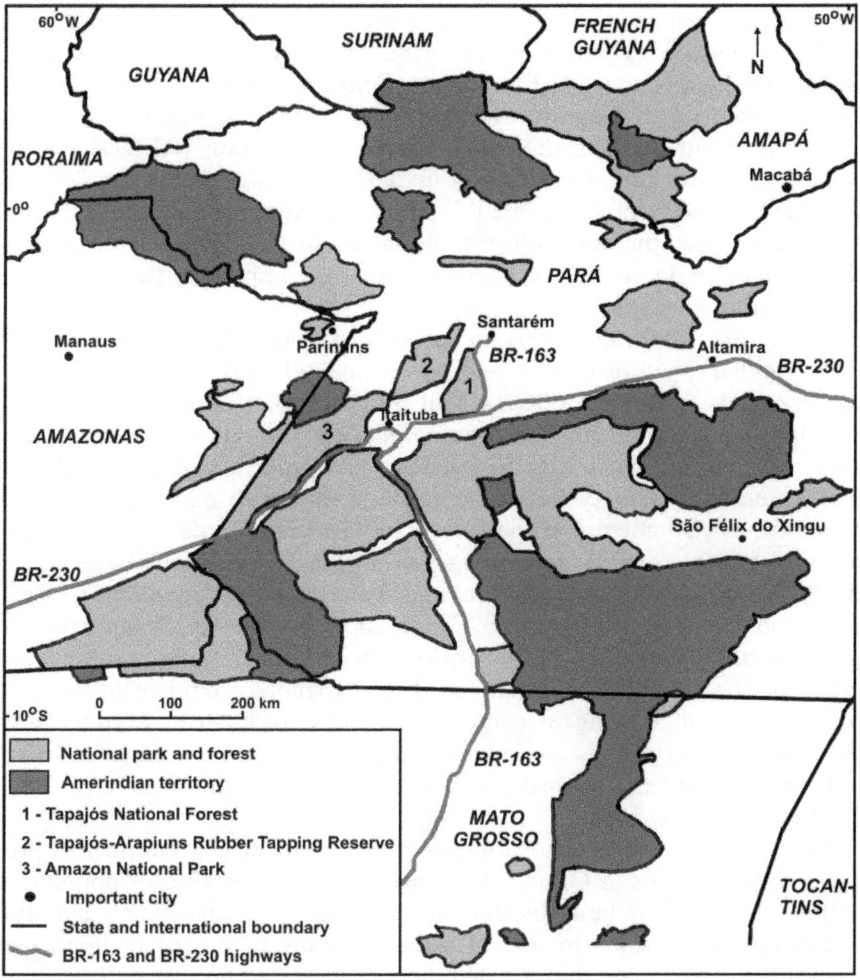

Fig. 1. Corridor of Conservation Units and Amerindian Territories Restricting Settlement off the Main Highways. *Source*: Adapted from IBAMA (2009).

In addition to setting aside enormous areas in conservation units another strategy for preservation relies on promoting alternative forms of income earned from non-timber forestry, handicrafts, and ecotourism coupled to transfer payments for environmental services that are meant to make the rain forest worth more intact than its potential in timber and commodity farm land. Such payments are usually made within a framework of emission reduction credits, which was established by a series of global agreements: the UN Framework Convention on Climate Change at the 1992 Rio Earth Summit, the 1997 Kyoto Protocol, and the Reducing Emissions from Deforestation and Forest Degradation (REDD) scheme ratified for forest protection in Bali in 2007 and reaffirmed in the Warsaw Climate Change Conference in 2013. The basic rationale until the 2008 economic crisis was for post-industrial countries to pay for nature conservation projects in developing countries in order to compensate their emissions in a global carbon trade-off system.[4]

During the 1990s and 2000s numerous sustainable development projects were funded by European-based governmental and non-governmental organizations within the PPG7 global program for reducing deforestation in tropical regions of the world. Some of these projects were successful in the short term but when it became evident that innovations did not diffuse to surrounding areas, conservation strategy shifted to making direct payments for environmental services to poor farmers in order to expand the reach of benefits over a larger geographical area. UN-REDD strategies in Brazil began with historic peasants and Amerindians living along rivers in preserved parts of the Amazon. Since 2007 the Forest Grant Program undertaken by the state government of Amazonas involving monthly payments of about US$33 per family has been expanded from benefiting 4,244 registered families living in conservation units to 8,090 in March, 2012. This transfer payment is associated to other community development projects meant to lift the poorest population of the state out of poverty (FAS, 2007, 2012; Viana, 2008).

In late 2011 the federal government began a similar new entitlement program called the Green Grant which targeted the same kind of population within conservation units located in other states of the North but also could include peasants in federal settler projects if located in important wilderness areas, like that treated here. The program began benefiting 18,000 families in the Amazon region and had the aim of reaching 73,000 rural families throughout Brazil by 2014 (Presidência da República, 2011; Viana, 2011). At the time of the last fieldwork in late 2013 the payment was about US$43 per month which could be combined with the national family support program, which varies from US$16 to US$133 per month depending on family size and income.

The two programs represent a vast improvement over previous neoliberal policy implemented in 1998, still in force outside conservation units, which places the burden of forest preservation on poor peasants by limiting productive land to 20% of the farm in forested zones of the Amazon.[5] However, in this study we will show that the income earned from non-timber products and the monthly Green Grant payments are a pittance in comparison to that earned from community-based logging or even from frontier farming and fishing. To some extent these shortcomings were overcome in the Tapajós National Forest so that the experience of this conservation unit points the way to how effectively engaging local actors created economically viable livelihoods despite the socially simplistic perceptions of administrators and environmental activists. The model of community-based forestry developed there is now being applied in the nearby Tapajós-Arapiuns Rubber Tapping Reserve from where it could be further replicated in other national forests, rubber tapping reserves, and conservation units permitting sustainable use. The opposite experience of the Amazon National Park is presented as a negative example of what should be avoided when dealing with new settlers from other regions of Brazil.

FROM RESISTING REMOVAL TO COMMUNITY-BASED FORESTRY IN THE TAPAJÓS NATIONAL FOREST AND THE TAPAJÓS-ARAPIUNS RUBBER TAPPING RESERVE

The Tapajós National Forest is the second oldest conservation unit in the Brazilian Amazon. It was originally set up in 1975 as a fig leaf for the vast destruction of forest provoked by the ambitious road building and colonization program of the military governments. The existence of this national forest spans a period of significant change in Brazilian environmental policy. It was originally operated by the Brazilian forest service IBDF (Instituto Brasileiro de Desenvolvimento Florestal) which emphasized commercial extraction of timber, then was transferred to the IBAMA when this agency was created in 1988 as the first real environmental protection agency in Brazil and today is run by the national parks service ICMBio. Until 2012 the National Forest occupied an area of 5,559 km^2 and due to favorable rainfall patterns has the highest biodiversity encountered in the Amazon. This together with close proximity to Santarém resulted in the national forest becoming one of the conservation units best studied by biophysical scientists in the Amazon (ICMBio, 2014).

The national forest is located in Belterra and Aveiro counties, about 50 kilometers south of Santarém, one of the most important regional cities of the Amazon with an urban population of 215,790 in 2010 (IBGE, 2010). The whole eastern side of the forest is accessed by the BR-163, a major highway connecting south to the rest of the Brazil. Today this highway is paved until Rurópolis, situated at the southern boundary of the conservation unit. Santarém is served by regular full-size jet service and has one of the most important regional ports, so that the location of the national forest has been crucial for the development of commercial activities and tourism there as well as for building partnerships with governmental and non-governmental organizations in order to access technical knowledge and public services.

The limits of the National Forest are clear: Belterra town to the north, the BR-163 to the east, the Tapajós River to the west, and the Gupari River to the south (close to the Transamazonian Highway, the BR-230). Due to this clarity of boundaries, the National Forest has suffered less illegal settlement than the more remote Amazon National Park treated below but the encroachment which occurred is shown to have important political implications. Most of the residents are riverine peasants who are situated along the Tapajós River where they have been living for centuries following the Conquest when a mestizo population (popularly called *caboclos*[6] in the Amazon) was created (more on this at the end of the section). The only significant incursion of settlers from other regions occurred along a secondary road linking the BR-163 to the riverine communities along the Tapajós River which gave rise to the community of São Jorge at the junction of this road with the main highway. These settlers have been living in this locality for more than 20 years and before 2011 allied themselves with the riverine population in the political mobilization which resulted in gaining land tenure. Another 30 farms and ranches are scattered along the western side the BR-163 clearly within the National Forest and are undergoing a slow process of being removed by the ICMBio through legal action involving expropriation.

The National Forest is zoned into areas according to the uses permitted (Fig. 2). An area along the Tapajós River and another area along the BR-163 in São Jorge were set aside for the resident population (1,438 families in 2006) which can use these areas for residence, small-scale farming, subsistence logging and hunting, and ecotourism. A logging management zone is situated along the BR-163 where sustainable forestry can be undertaken. More restrictions are made in a zone of preservation and non-logging forest management where only latex, native fruit, nuts and honey can be collected and ecotourism developed by the local population. Finally, in the zone of complete preservation only authorized research and ecotourism can be practiced. A buffer zone of 10 km exists around the

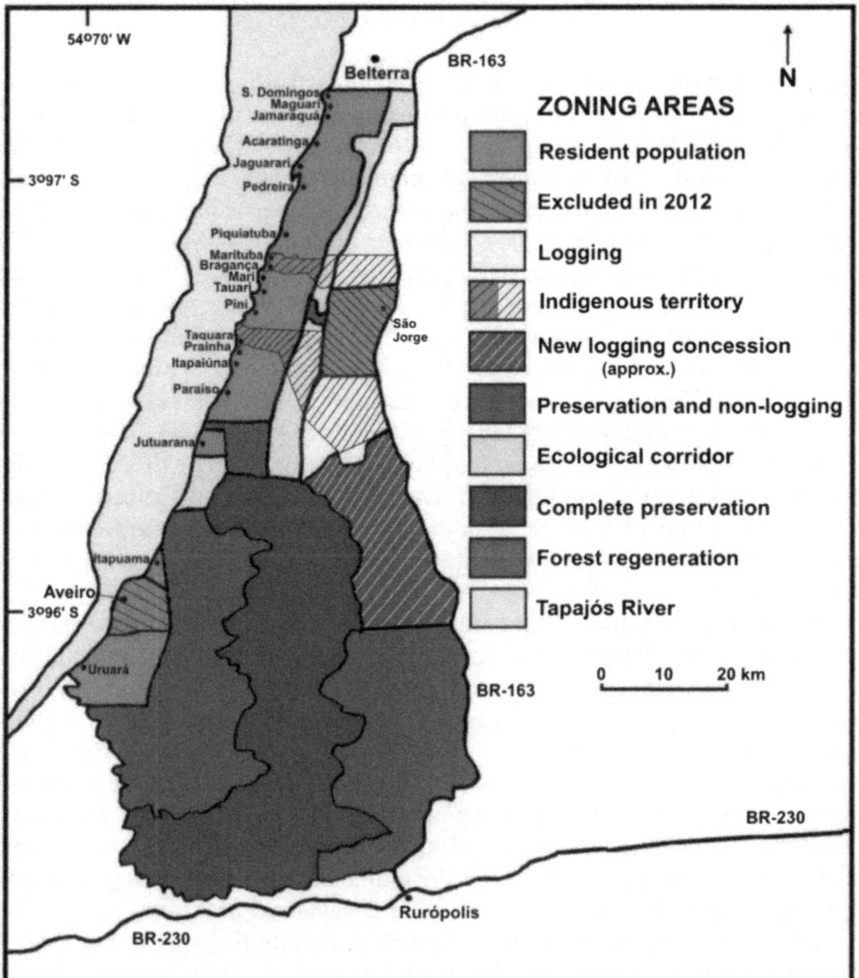

Fig. 2. Land Use Zoning in the Tapajós National Forest. *Source*: Adapted from IBAMA (2005b), ISA (2014).

National Forest where in theory only sustainable activities can be undertaken but in fact typical unsustainable frontier farming prevails along the roads which pass along the eastern and southern boundaries of the National Forest.

Community-Based Commercial Logging

Community-based commercial forestry is undertaken in the logging management zone. A sector rotation system was initiated in 2009 in an area of 32,222 hectares located along the BR-163 in the northeastern part of the National Forest. The management plan permitted logging one 1,000-hectare sector per year which after extracting timber is left undisturbed so that forest re-growth occurs over a 30-year period. The system is based on decades of empirical research undertaken by the EMBRAPA Floresta, the forest-research branch of the Ministry of Agriculture, which used the national forest as a laboratory for studying the impact of different forestry systems. The system developed is based on recommended practices of the International Tropical Timber Organization (ITTO) for sustainable forestry in which reduced-impact logging methods are used. The local people's familiarity with the flora, fauna, and landscape is mobilized to identify biodiversity and terrain conditions during the different operational stages, constituting a perfect example of uniting scientific and local knowledge, frequently mentioned in environmental rhetoric but seldom encountered in actual practice.

Sustainable forestry is far more complex than conventional timber extraction in the Amazon and employs pre-logging and logging activities which require meticulous work and detailed knowledge of the forest. Pre-logging planning activities consist of delimiting logging sectors, undertaking forest inventory (botanical identification of tree types), projecting a road network and storage facilities in the sector to be logged, determining the direction of felling and dragging trees in order to avoid damaging forest, choosing trees to cut according to local availability of species and protection of rare and endangered flora and fauna. Logging activities can begin only after the annual logging plan is reviewed and approved by the ICMBio in Brasília and this can cause delays in initiating operations due to slow bureaucratic processing. Roads and storage facilities are built in the sector to be logged, trees are tested to avoid cutting hallow trunks, selective cutting is made to avoid reducing biodiversity, stumps are left to guarantee forest regeneration, vines are cut to minimize damage to other trees, and cutting trees with bird and animal nests is avoided (Fig. 3).

Today this forestry system in the Tapajós National Forest is undertaken by the local rural population and not commercial firms because of a contentious 40-year struggle for rights in which strengthening collective goals produced the social capital necessary for undertaking community-based

Fig. 3. Planned Tree Felling to Limit Damage to the Forest. *Source*: Authors' Fieldwork (2013).

forest management. There were two dimensions to this struggle, one for land and the other for the right to exclusively use forest products.

The struggle for the right to land pitted the Tapajós National Forest administration, the IBDF and later the IBAMA on one side against the Brazilian Land Office (INCRA) and the local population on the other. In 1975 when the National Forest was created it overlapped with existing INCRA settlement areas and riverine peasant areas. At the time, the National Forest administrators tried to implement the conventional U.S. style policy of removing the people resident in 18 communities in order to reinvent a pristine "nature" which never existed. The riverine population had a common social background of kinship and community ties so that it was politically better organized and more active in pressuring the IBAMA for land tenure rights than settlers along roads.

In 1980 the riverine communities unilaterally set their land use area to 10 km inland from the river and in 1983 they paralyzed IBDF land survey work meant to remove them. In 1990 IBAMA made a forest inventory and the population mobilized again to resist eviction. The Santarém county government came to their aid and in 1992 proposed to exclude the 10 km area along the Tapajós River from the National Forest but the IBAMA rejected this. In 1996 a plebiscite was approved excluding the community area but again the IBAMA did not abide by the result of the plebiscite. In 2002 a Local Land Tenure Commission was created to negotiate a concession within the National Forest which another plebiscite approved in 2003. Finally, in 2010 an agreement was signed conceding land use for 20 years. This was possible because in the meantime the environmental agencies had changed their mentality and went from being adversaries to becoming allies in sustainable development projects.

At different points in the struggle considerable potential for open violence existed. When talking with an apparently peaceable old man he pointed to a wooded area on the edge of the community and said that an armed conflict almost took place there about 20 years before. In a meeting held in the community on one night at a particularly tense point of negotiating with the national forest authorities the alliance of communities mustered a force of 80 armed men who hid in the woods waiting to enter into action depending on the behavior of the federal police who accompanied the negotiators. Nothing eventful happened which the old man thought to be for the better, "because otherwise it would have been war!"

The other dimension of the local people's struggle involved the right to use forest resources, which pitted the IBAMA, researchers, and logging firms against the local population. In the 1980s, official forestry experiments were undertaken in the form of joint ventures between the IBDF and private logging firms which used local people as workers. The extensive scientific biophysical research undertaken in the national forest also employed local labor. In both kinds of work the local people gained considerable experience in forestry activities. However, as the experimental forestry area overlapped the area of the riverine communities, it represented another source of conflict which led to demands for excluding outside private firms in favor of community forestry management for local benefit.

The social movements and political organization of the local population which arose in their struggle for land and resource rights involved a long process of community organization, culminating in the Tapajós National Forest-wide cooperative set up to undertake sustainable commercial logging. During the 1970s, and much like occurred in other parts of riverine Amazonia, the progressive wing of the Catholic Church induced community organization in specific localities along the Tapajós River which was meant to arbitrate conflict involving fishing and farming rights between members and to resist incursions by commercial fishing boats as well as to pressure local government for public education and health services.[7] This organization was also used to channel land tenure resistance and to form alliances with the INCRA and municipal government against the federal forestry and environmental agencies. Between 1997 and 2005 legal recognition of the community organizations in the form of associations was achieved which added strength to their struggle for land and forestry rights. The associations became formal entities of collective representation for negotiating directly with public institutions which enabled participation in productive projects located in communities and permitted building wider alliances with local, national, and international governmental and non-governmental organizations. At the same time, three inter-community associations (AITA, ASMIPRUT, APRUSANTA) were set up to represent different groups of communities. The three then came together in order to undertake collective logging and other forestry projects as well as to provide technical training for members via projects with NGOs and the IBAMA.

Based on this wider organization the COOMFLONA cooperative (Cooperativa Mista FLONA Tapajós Verde) was set up in 2005 as a requisite for gaining the exclusive right to extract and sell lumber and other non-timber forest products from the National Forest via a community forest management system and a coop marketing structure. The objectives of the cooperative are to operate community commercial forestry in order to create work and generate income in a part of the Amazon characterized by subsistence production and high underemployment. As the cooperative is managed exclusively by people from the national forest, local perceptions prevail and logging was defined as the main commercial activity because it is the most highly valued regional product. Non-logging products are undertaken by individuals and community associations with the help of the cooperative and the cooperative markets these products. The cooperative has since obtained green forestry certification which allows it to sell to eco-markets. It also provides the institutional means for undertaking local and distant technical and administrative training in logging and non-logging production as well as in cooperative administration and legislation, so building advanced human capital.

At the time of the research undertaken in 2010 the cooperative logging operation had a work force of eight administrative and accounting personnel (with five people from Santarém contracted), 55 field workers during the cutting period of six months per year and a smaller group doing pre-logging activities during the rest of the year. To spread the benefits field workers must be from different communities, only one worker per family and are selected by the local associations. From net proceeds, the cooperative distributes income and benefits in the following fashion: 45% goes to investment and operating capital, 20% for worker productivity, 15% for a contingency fund, 15% for community buildings and activities, and 5% for community health care. Payment received in logging operations that year varied from US$2,333 to US$5,667 per worker plus 20% more in productivity payments to field workers.

The first year of operations was 2010 when 700 hectares were selectively logged and the next year this expanded to 1,000 hectares so that another 15 loggers were hired. Then in 2012 an institutional conflict with the Brazilian Amerindian protection agency FUNAI (Fundação do Índio) resulted in the loss of half of their logging area but the cooperative was compensated with another area more than twice as large further south in the National Forest so that beginning 2014 the logged area was doubled to 2,000 hectares a year and the work force increased accordingly. With a view to creating more work and income, the cooperative has plans to build a sawmill in order to aggregate greater value than that earned from selling brute logs and this will create jobs for another 50 workers. With this, families with members employed by the cooperative will pass from 5% to 14% of the total number of families living in the National Forest.

Non-Timber Forest Products, Crafts and Ecotourism

Induced by environmental agencies and foreign governmental and non-governmental organizations who favor low-carbon emission non-timber forestry over logging, non-timber activities, such as collecting Brazil nuts, natural oils, and tree seeds, extracting latex, making wood handicrafts and furniture, and undertaking ecotourism, have been developed by associations and individuals in the community area of the National Forest. These activities are practiced in different communities spread out over the National Forest, which could have the potential for including more people of all ages and gender while lumber jacks are younger men because logging demands considerable strength and physical stamina (Table 2). The activities are based on local knowledge and skills but are relatively small-scale and do not generate much income. Due to their environmental and cultural appeal these activities are highly encouraged by the NGOs but are not given priority by the cooperative which has a better sense of what is commercially viable.

Furniture making could be promising because it represents a way of creating employment and aggregating value. Santarém is a relatively large

Table 2. Spatial Distribution of Non-Logging Activities by Community in the Tapajós National Forest.

Activity	Community
Furniture, window, doors	São Domingos, Paraiso, Tauaí
Decorative carvings	Pini
Wood inlay handicrafts and furniture	Pedreira
Honey	Maguari
Latex-leather handicrafts	São Domingos, Maguari
Natural oils	Maguari
Brazil nut collecting	All communities
Ecotourism	São Domingos, Maguari, Jamaraquá, Acaratinga, Tauarí, Paraíso, Itapuana

Source: Authors' Fieldwork (2010, 2013) and IBAMA (2005b).

urban market by the restricted standards of the Amazon and the city is not far away so that the carbon footprint generated would be minimal. One workshop was set up with a grant from the Kreditanstalt für Wiederaufban but only made beds for the barracks of the coop logging workers and not much else. Another workshop builds more refined furniture involving wood inlay and the products are sold in the coop headquarters in Santarém.

Ecotourism is another activity favored by environmentalists but the scale is quite limited and is only undertaken in Maguari and São Domingos, the two most northerly communities of the National Forest which are accessible by road. Ecotourism consists of guided hiking in the forest which costs about US$23 per tourist. In one community six guides received about 40 visitors in 2010 and seven guides 60 visitors in the second community, for an average annual income per guide of only US$152 and US$200, respectively. This income is highly seasonal, when hikes are organized during three months of the year when it is not raining.

Bee keeping is another alternative activity which can produce a respectable income and one farmer earned US$6,800 in 2010 but honey has a limited market in the Amazon. Another association produces oils extracted from cumaru, andiroba, copaíba, and pequiá trees growing in the forest and also planted nearby houses in the case of andiroba. In 1999 there were 22 men and women involved but this fell to 11 in 2009 after a number of problems with unscrupulous foreign buyers and restrictions to exporting were imposed by the IBAMA. The first sale abroad involved 250 liters sold at US$30 a liter which yielded US$341 for each participant but the buyer of the second shipment reneged payment claiming that the oil arrived spoilt. With this, interest in producing oil declined and today the 11 producers who remain active earn about US$500 a year each from this line of work.

Another association involving 19 families produces latex products, which is perhaps the most famous of alternative activities promoted by NGOs in the Amazon. Bags and flip-flops are made at a cost of about half the final price received, US$17 and US$14, respectively. Against the price received by the producer, a considerable surcharge for authenticity is made by the middleman selling to eco-consumers in Europe, where industrially made Brazilian flop-flops of the Havaianas brand command a price of 20 euros. Again problems with unscrupulous buyers occurred resulting in non-payment which nearly caused the breakup of the association. A more serious problem occurs with the quality of the bags and flip-flops made from latex which tend to be sticky. One local producer has overcome these technical and marketing problems and judging by the quality of his house, furniture, and appliances, he makes a significant income even if exact data on yearly income was not available (Fig. 4). This was his only source of income and like the other interviewed residents of the National Forest he receives no remittances from adult children living in urban areas where the minimum wage barely covers basic necessities.

Fig. 4. The House of an Eco-Crafter Family Built with Durable Materials which must be Bought. *Source*: Authors' Fieldwork (2010).

Multi-Functional Livelihoods for All?

As field research in 2010 was biased toward families living in the National Forest who were engaged in new green activities located in specific communities, fieldwork undertaken in 2013 systematically investigated rural livelihoods of typical families living inside and outside of the Tapajós National Forest and the Tapajós-Aripiuns Rubber Tapping Reserve. As income earned by lumber jacks is high by local standards and does not benefit more than a fraction of the families living in these conservation units the object of the new research was to ascertain the socio-environmental impacts of the limitations to farming and rural livelihood imposed by conservation unit authorities vis-à-vis farmer families living outside the units who do not suffer the same restrictions. In addition, research over a wider area detected considerable spatial variation in activities and income within the conservation units.

The two conservation units occupy opposite banks of the Tapajós River which historically was the focus of human habitation in the region. The Tapajós is a crystalline river with no significant suspended sediments so that farm land located on its margin has low fertility due to high sand content in the soil. The western side of the river has flat bluffs and in the past was more densely populated by Amerindians who practiced in-field mulching and agroforestry systems which artificially created richer soils (Denevan, 1996). The eastern bank of the river has low-laying sandy beaches and marsh areas located along the river and rugged high-rising bluffs above the river so that this side of the Tapajós has less potential for farming and habitation. During the 20th century the western bank, now located in the Tapajós-Arupiuns Rubber Tapping Reserve, had more rubber production and a denser population than that present on the eastern side of the river in the Tapajós National Forest. In addition to these natural limitations, farmers within the conservation units are not allowed to plant more than one hectare in the National Forest and not more than two hectares in the Rubber Tapping Reserve. In both they cannot plant pasture for cattle, which is the main economic activity of the Amazon. Many farmers reported that in the past they planted larger crop areas and some raised cattle but suffered pressure from the National Forest administrators to reduce these activities. Consequently, most farmers only plant about a quarter to a half of hectare of bitter manioc, beans, and corn for self-provisioning. Some farmers sell part of their manioc but two-thirds of the typical families of the conservation units earned no income at all from cropping.

As a result, most residents of the National Forest are dependent on government transfer payments in the form of retirement pensions and family support programs for monetary income. Lack of opportunity has induced young people to migrate to Santarém while the elderly who stay behind live off pensions and head 48% of the households. Of the children over 18 years old, only 38% still live in the National Forest while 58% left to live in Santarém, Manaus, and other regional cities. In 2012 the federal government began conceding Green Grant payments in the study area but first to the COOMFLONA workers and in communities located in the north of the National Forest where higher incomes are earned. In late 2013 residents in the south of conservation unit had still not received any environmental payments despite the fact these communities are among the most impoverished encountered by the researchers in decades of experience in the Amazon (Fig. 5).

The livelihood strategies of three COOMFLONA workers interviewed in 2013 were found to be diversified to the point that they can be considered multi-functional. Not only did the workers have higher monetary income earned from the cooperative during the dry season but they also invested part of this income in different types of farming which are practiced during the rainy season. The dual seasonality of activities in turn avoids the most common problem with building multi-functional

Fig. 5. The House of a Family in an Impoverished Community in the South of the Tapajós National Forest. *Source*: Authors' Fieldwork (2013).

livelihoods, namely conflicting work schedules or long working hours. Combining logging and farming activities at different times of the year also avoids seasonal underemployment typical of the Amazon. The diversity of farm activities included planting a hectare of manioc which can be harvested slowly at the producer's convenience, raising chickens and a few steers of cattle and rubber tapping. One man in the National Forest planted a field 3 km inland located up on the plateau at a point deeper into the forest where land is twice as fertile but working land in primary forest is not exactly what officials want. Normally farmers of the National Forest only plant in areas of secondary forest located up to 500 meters inland. The COOMFLONA workers are also less dependent on family support programs and because they work in the cooperative all receive Green Grants. Summing all the different forms of income, the total earned by COOMFLONA workers was more than three times that of typical farmers within and outside the conservation units and was more than twice the minimum wage (Table 3).

Riverine farmers living outside of the conservation units on average earned US$4,352 in 2013, which is only marginally better than the average income of US$4,207 obtained by small farmers living inside the conservation units. This happens because few commercial products withstand the long boat trip to Santarém or Itaituba. Consequently, the farmers on

Table 3. Variation in Annual Income according to Source for Farmer Families within and outside the Tapajós Conservation Units (US$2013[a]).

Rural Activities	Small Farmer Outside C.U. ($n = 8$)[b]	Small Farmer Inside C.U. ($n = 25$)[b]	COOMFLONA Worker-Farmer Inside C.U. ($n = 3$)
Cropping	0–3,251	0–3,695	2,771–4,064
Chickens, pigs, sheep	0	0–544	0–95
Cattle	0–67	0	0–759
Latex, tree oil, etc.	0	0–897	0–1,773
Fishing	0–3,399	0–1,773	0
Wage work off land	0–4,274	0–3,369	7,044–13,655
Pension, family support	0–8,039	0–8,887	597–1,714
Green grant	0	0–591	591
Total	1,212–11,424	413–11,217	11,099–21,798

Source: Authors' Fieldwork (2013).
[a]US$ = R$2.03.
[b]Non-farmers excluded.

average only plant close to one hectare of manioc, beans, and maize some of which is sent to these markets. Without the conservation restrictions they deforest more land to plant between 3 and 20 hectares of pasture for raising cattle. However, pasture planted on river slopes causes erosion. They also undertake more commercial fishing which can deplete stocks to point that some communities forbid outsiders from entering their areas. With these activities farmers diversify their sources of income and are less dependent on government transfer payments but the resulting combination of activities hardly qualifies as a sustainable multi-functional livelihood which promotes environmental conservation and social inclusion. Half of the children over 18 years of age remain in the riverine communities, 37.5% left to live in regional cities, and 12.5% moved to areas of settlement along roads located close to cities.

Income is also low for smallholders who plant in poor soils along the roads and are located far from Santarém, where interviewed farmers (n = 6) only earned US$1,118 from rural activities and another US$1,345 from transfer payments in 2008. Crop infestations such as *Moniliophtora perniciosa* in cocoa and cupuaçu trees caused the collapse of agroforestry systems which had been the great hope for sustainable farming systems along roads of the Amazon when first introduced in the late 1990s.[8] With few monetary alternatives families have moved to the cities of Itaituba and Santarém. One man mentioned that only he and another farmer are the last of the original settlers still active along a whole feeder route. Açaí and fruit farming still generates decent income, US$3,334 in 2007 (US$ = R$1.77), but until recently was only viable for interviewed farmers located near Santarém (n = 5) where rural electricity for freezing pulp was available. Similarly, intensive chicken farming for the Santarém market is a good option along paved roads and one interviewed farmer reported earning an annual income of US$21,983 in 2009 (US$ = R$1.74) but this activity demands long hours of work which leaves little time for other rural activities or even leisure.

These regional trends demonstrate that the local people in charge of the community-based logging cooperative when faced with the environmental and logistical limitations outlined above were right to concentrate on extracting timber. Over time riverine farmers have undertaken many of the non-timber activities promoted by governmental and non-governmental environmental groups but eventually gave them up when they ceased generating sufficient income. Latex is a case in point. Demand has picked up for producing ecologically friendly tires and condoms so that latex is being extracted again for the first time in decades. Demand for latex tires for race cars and condoms for allergic users may not create a significant export market but the opening of a factory in Manaus in 2012 where latex is used to produce tires for motorcycles and bicycles for this booming regional market is more promising. Income from latex is still quite modest for interviewed farmers, representing respectively 9% of the income of farmers who live outside the conservation units, 3% for those inside, and 4% for COOMFLONA workers. This is little when compared to income earned from other rural activities and is a shadow of what was made during the rubber boom of the past.

Simplistic Ethnic Models of Cultural Complexity Dividing the Movement

New trends in ethnic and regional identity threaten the political cohesion of the conservation unit-wide community movements which sustain the forestry activities. The conflict involving FUNAI cited above reveals political problems simmering in the Amazon as Brazil increasingly adopts U.S. style affirmative action policies of dubious applicability in the country. Brazil is a well-known melting pot of Amerindian, African, European, and Asian ethnic groups which created a highly inequitable multicultural society. Racial discrimination no doubt exists but in the absence of clear-cut markers separating ethnic groups it is difficult to revert through affirmative action. As part of affirmative action meant to correct past injustice to Amerindians, extensive indigenous territories have been created, public services improved and entrance quotas created to public universities. The census now allows individuals to self-declare their racial group and many inhabitants of the Amazon are opting for Amerindian instead of mestizo (*pardo*, literally brown) because they think that in indigenous territories they obtain collective title to the land, not just a temporary 20-year concession like in conservation units, are not subject to strict environmental control of their activities and receive better health care and educational services from FUNAI than from county government.

The Tapajós National Forest has three communities of Munduruku descent who migrated to the area in the past.[9] Today they number about 500 individuals and live in Bragança, Marituba, and Taquara (ICMBio, 2014; Vaz Filho, 2011). As part of a highly successful region-wide political movement of native peoples of the Amazon, the three communities declared themselves to be Amerindians in order to receive an indigenous territory. In this process they now legally live in a "village" under the jurisdiction of FUNAI so marking their separation from the "communities" of the riverine peasants. For the latter, "community" is a term highly charged with the symbolism of the collective political movement for land, resources, and public services. The Amerindians not only claimed the 10 km area adjacent to their villages along the Tapajós River like the riverine peasants but also the whole cooperative logging area conceded to all of the communities. This area is located along the BR-163 on the other side of the National Forest but was claimed to be ancestral hunting grounds, which is preposterous given its location up to 30 km inland and the elongated shape of the area claimed. The ICMBio defused the issue with the territorial compromise described above and the COOMFLONA even agreed to train the self-defined Amerindians how to undertake collective forestry in the area they gained in the north of the National Forest. However, the logging area can only be granted as a concession because commercial timber extraction is officially prohibited in indigenous territories. The prohibition exists because it has been common for loggers to extract timber illegally from indigenous territories with or without the permission of indigenous leaders who may be threatened or bribed into agreeing.

The villages also received indigenous schools, first provided by the county government to be replaced later by official FUNAI schools (Vaz Filho, 2011). These will re-teach their ancestral language and customs and one study cited Hobsbawn's concept of re-invented tradition approvingly to justify this (Costa, Vieira, Lima, & Mileo, 2013), even if Hobsbawn was ambivalent about this phenomenon (1969, p. 170). Relearning one's culture is great for self-esteem but can take strange twists and turns in which visits to the villages to see "tribal dancing" are now billed as a tourist attraction by private-hire cruise boats in Santarém. This kind of performance for tourists takes place near Manaus and is pathetically inauthentic for anyone who has actually seen Amerindian dancing in a tribal setting. The performances border on fraud when undeclared individuals of American descent are dressed up as "natives" to dupe the tourists (see Hoefle, 2014 for more on tourism in the Amazon).

The problem with all of this is that the people of the other communities of the National Forest are virtually indistinguishable from the self-defined Amerindians. Indeed, one of the authors did fieldwork with "really tribal" Amerindians (Yanomamös and Ye'cuanas of northwest Roraima state) 40 years ago and was amazed how a man from a neighboring community displayed behavior nearly identical to that observed in 1975. While listening to his complaints about the claims of the Amerindians, the author noticed that in addition to having pronounced Amerindian ancestry the man was laying on his side in a traditional homemade cotton fish-net hammock and was not laying belly up diagonally across a manufactured cotton hammock as would be the case for a non-Amerindian. The whole scene reminded him of numerous similar conservations with an old Ye'cuana friend years ago.

However, much more than re-invented tradition is at stake. Fault lines exist between academic specialties of Amerindian, peasant, and urban studies in Brazil, which have tangible political consequences for local people. Based on questionable models of the pristine cultural purity involving Weberian ideal social types, complex cultural landscapes are distilled/essentialized into segregationist entities.[10] In the case of indigenous territories only Amerindians can be present and the "non-indigenous" must leave, which causes conflict within mestizo communities and even within families during the process of setting up new territories. One facile view of this is that a self-declared Amerindian is a mestizo (*caboclo*) who has become politicized (Peixoto, Arenz, & Figueiredo, 2012) suggesting that the latter is politically passive which we saw above could not be further from the truth.

During the 1990s a movement of the riverine population of the Arapiuns River and the western side of the Tapajós River successfully pressured the IBAMA to set up a rubber tapping reserve, which was created in 1998 in an area of 6,775 km^2. This movement mirrored that of the residents of the Tapajós National Forest but involved resistance to encroachment by commercial logging firms and not to the IBAMA.[11] Vaz Filho (2010) did a detailed study of identity politics of the movement in the Rubber Tapping Reserve and shows that until 2002 local people were able to support both the reserve and the creation of indigenous territories. However, beginning 2002 a new center-left federal government came to power in Brazil which proved to be extremely corrupt and wrong-headedly progressive. High-level corruption in the environmental agencies involving a large grant from the Finnish government to implement a community-based forestry system in the Rubber Tapping Reserve similar to that in the Tapajós National Forest drove a wedge between residents polarizing them into those in favor of and those against indigenous territories within the conservation unit.

This notwithstanding, when Vaz dismisses those in favor of community-based logging as "economistic" he is patronizing at best. We are not dealing with isolated Amerindians living in a tribal economy. The Tapajós River has been a major economic thoroughfare for centuries and with the rubber boom from the late 19th Century onward this area received a large influx of Northeasterners (themselves with considerable Amerindian ancestry) who intermarried with the locals. Today the descendents of these people are not tribal Amerindians who do not use money and are not subsistence-minded peasants who are satisfied with just affirming their ethnic pride but also want rights to land and natural resources with which to earn enough income to live with dignity. That is what the world-wide forest tenure movement is all about.

Simplistic models of segregated ethnic grouping have given rise to the absurd situation whereby families in communities located in the Rubber Tapping Reserve declare themselves indigenous and then pressure the other residents of the communities to declare themselves Amerindian or leave. One of the leaders of the Rubber Tapping Reserve, dismissed as "economistic" by Vaz, is of Amerindian descent, recognizes the fact, but does not accept his brother pressuring him to declare himself indigenous because he feels that this is dividing their wider political movement. Indeed, the man holds the view that many of the individuals declaring themselves indigenous are opportunists who want an easy route to land and better public services. Who is being politically conservative here? This whole situation brings to mind Frake's (2014) recent criticism of NGOs and the invention of The Indigenous, particularly that concerning the problem of validating one's status as a qualified member of an artificial category that was originally created to oppress native peoples of the world.

A plebiscite held in 2011, which resulted in the exclusion of the area of São Jorge community and the county seat of Aveiro from the National Forest, further complicated political identity in the study area. An urban zone located within the National Forest was always a problem, particularly because many of the townspeople operate cattle ranches nearby. The people of São Jorge are settlers who originally came from the deforested state of Maranhão so that their farming practice also was out of synch with what is permitted in a conservation unit. In the plebiscite the residents of the National Forest were given the choice of staying in the conservation unit and receiving Green Grants or having their area excluded, which was what the people in the two areas voted for and in 2012 69 km^2 were removed from the National Forest. The people of São Jorge had participated in the decades-old political movement for land, gained it and had nothing else in common with the riverine peasants. The latter understood this and said that the settlers of São Jorge voted the way they did because they are not "traditionals" (historic peasants or Amerindians). Even the ICMBio authorities were relieved to see them go because inspecting and fining farmers in the two areas always created considerable tension.

UNSUSTAINABLE FRONTIER FARMING IN AND NEAR THE AMAZON NATIONAL PARK

The case of poor peasants invading the Amazon National Park does not involve the cultural and political complexities of the other two conservation units treated here but is typical of the challenges to environmental policy involving settlers from other regions, who squat on public land located along roads. The park was created in 1973 and is the oldest full conservation unit in the Amazon. It was originally created with an area of 9,459 km^2, expanded to 11,120 km^2 in 2006 and then was reduced to 10,707 km^2 when land was granted to the peasant invaders in 2012 (Fig. 6). The

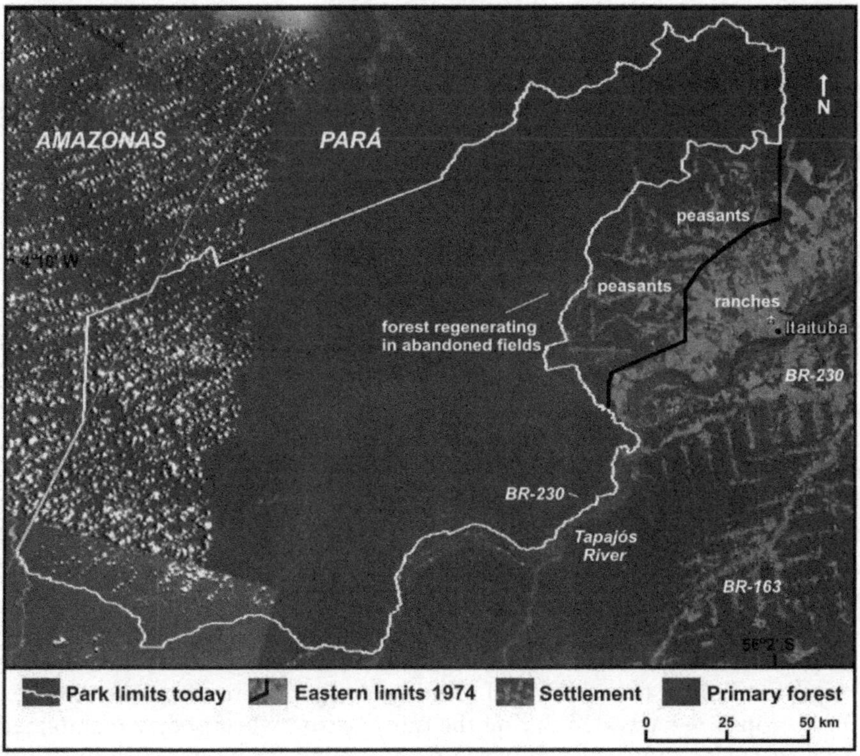

Fig. 6. Current limits of the Amazon National Park (white line) and Previous Eastern Boundary (black line) before Excluding Peasant Settlements.
Source: Adapted from ICMBio-Google Earth (2014).

conservation unit is located in western Itaituba county as well as parts of Aveiro county and Maués county (in Amazonas state).

The only natural limit of the Park is the Tapajós River to the south where the Transamazonian Highway passes along the edge of the Park. As there is no clear limit visible to the east, after 1990 settlers entered the buffer area and the Park itself when they occupied the areas behind a mega-cattle rancher, who claims hundreds of thousands of hectares of land originally destined for INCRA smallholder settlement projects. The rancher did not oppose the invasion because he could claim that the peasants occupied what would be his forest reserve so freeing him of this encumbrance. He also supported their cause against the National Park politically and was the runner up in the election for mayor in 2008, which shows that settlers and ranchers can be political allies today and not just opponents as was the case in pre-1988 Amazonia. Indeed, it has become common practice for large holders to place smallholder clients on land adjacent to or within the land they claim with the intention of complicating the tenure situation in order to make it difficult for the environmental agencies to take action.

Despite being located along the Transamazonian Highway, settlement in this part of Pará was slowed by the Itaituba gold rush so that an effective expanding frontier of new settlers was only established after 1990. Gold was first discovered on the tributaries of the Tapajós River in 1958 and over the next three decades prospectors swarmed to the area establishing more than 300 camps and 200 dirt air strips to access remote areas where gold was to be found blasting the river bank with water pressure equipment. The environmental impacts were degradation of river banks, silting of creeks, and mercury pollution in the rivers (Mathis, 1998; Rabelo, 2006). The immediate impact on population movements besides attracting outsiders directly to the camps was to empty the planned colonization projects being established at that time along the Transamazonian (BR-230) and the Cuiabá-Santarém (BR-163) highways, which delayed the process of settling western Pará by decades (Hoefle, 2013).

When the gold rush collapsed in 1990, some ex-prospectors became frontier peasants on the road heads of Itaituba county. Today they suffer the typical limitations to commercial farming along most roads in the Amazon where crops are harvested during the rainy season when poorly maintained unpaved feeder routes and unpaved main roads become mired in mud and traffic becomes impassable.[12] Some of the frontier farmers try to raise cattle prime for sale after the rainy season but a standard 100-hectare lot, of which since 1998 legally only 20 hectares can be cleared for use, does not furnish enough pasture to support a herd large enough to generate a decent living. As a result, farmers crop beans, corn, manioc and fruit trees for subsistence and rice for the market. Pigs and poultry are raised mainly for self-provisioning and some cattle are sold in the limited urban markets of western Pará, where cities are few and far between. However, the main cash activity, rice production, is being curtailed due to the appearance of a fungus which kills the crop, a perennial problem when commercial-scale cropping is attempted in the Amazon, so that little monetary income is earned from farming.

These problems can be highlighted by comparing small farmer activities out on the road heads, located from 50 to 70 km distance from Itaituba, with those of farmers located in areas of older settlement along the BR-230 just east of the county seat. These farmers use reasonably maintained unpaved roads and have access to a nearby market in Itaituba with 70,682 urban consumers (IBGE, 2010). With average farm size of 126 hectares they own over 50% more land than farmers on the road heads (Table 4). With sub-standard farms of 84 hectares on average, producers located further away from Itaituba experience monetary difficulties and end up selling parcels of their land. Farms near Itaituba have more land in cropping which reflects better prospects for selling produce but both types of small farmer have very little land in fallow, which in order to recover soil fertility naturally should be at least five times that of the land in crops, if not eight times (the world average for minimal fallowing for sustainable slash-and-burn cropping in tropical regions according to Boserup, 1965; Simmons, 1989).

Lack of sustainability in frontier farming causes high spatial mobility as farmers move on when the natural fertility of the poor rain forest soils is exhausted. Most farmers hail from western Maranhão state where forests were devastated by this process after 1960 and eastern Pará after 1970. Fifty-seven percent of the interviewed farmers on the road heads have moved five to eight times during their lifetime and another 16% from nine to fifteen times. Many of the moves took place due to the high mobility of gold prospectors but exhausting the soil through lack of fallowing was

Table 4. Land Use on Farms in Itaituba County (hectares).

Location	Crops	Pasture	Fallow	Forest	Total
Distant ($n = 40$)	4.6	7.2	3.3	69.5	84.6
Nearby ($n = 8$)	6.4	7.2	9.9	103.3	126.8

Source: Authors' Fieldwork (2008).

responsible for high mobility over the last 25 years. Farmers located near Itaituba are relatively successful and have lived on their present farm for an average of fifteen and an half years and before this 83% had moved only four or less times during their lifetime.

Mobility is also related to low incomes. As happens in much of the Amazon, there are a relatively large number of farmers trying to sell to the limited urban markets of the region so that not much produce is actually sold. On average farmers located near Itaituba may earn twice the average income of distant farmers, US$4,276 versus US$1,984 versus in 2008, but farm income for both groups is low. The high dependence on government family support programs for monetary income, which represents more than half of the income of road head farmers, is a sign of the difficulties involved with generating income from frontier farming (Table 5).

What the farmers lack in monetary income they compensate in political pressure on county officials to provide basic education and health services. Most farmers interviewed are from Maranhão and have the same life path through the gold camps, which means that they have similar social backgrounds and this facilitates political union. Today, they live in legally constituted communities and even organize group projects in order to finance better housing than that usually found along the cutting edge of the Brazilian frontier (Fig. 7).

This community organization also helped them to mount a resistance movement to avoid being evicted from lands located within the National Park. Inter-community alliance occurred through the local Farm Worker Union which reached a compromise with the INCRA and the IBAMA, whereby the boundary of the Park was shifted twenty kilometers westward

Table 5. Variation in Annual Income according to Source for Farmers Families in Itaituba County (US$2008[a]).

Rural Activities	Small Farmer near Itaituba	Small Farmer near A.N.P.	Crop Worker near A.N.P.	Ranch Worker near A.N.P.
Cropping	257–13,311	0–5,846	0	0
Chickens, pigs	0–3,086	0–2,857	0	0
Cattle	0–2,857	0–3,314	0	0
Wage work off land	0	0–343	343–514	2,143–3,429
Family support	0–2,902	0–6,171	446–837	0
Total	2,044–16,647	0–6,682	789–1,351	2,143–3,429

Source: Authors' Fieldwork (2008).
[a]US$1 = R$1.75.

Fig. 7. House in Community Located Near the Amazon National Park. *Source*: Authors' Fieldwork (2008).

to two rivers in order to clearly mark the limits to the Park and the farmers beyond that point pulled back to the part conceded for settlement. The role of the local Farm Worker Union was unusual in the process. These unions in Brazil usually represent workers and smallholders, while medium and large farmers have separate associations. In neighboring Amazonas state, farm worker unions have little political importance, usually limiting their activities to processing requests for rural retirement, while in the land conflicts of Pará state they have an important political role in representing smallholders (Bicalho & Hoefle, 2010). This arrangement redrawing the eastern limits of the Park was formalized in early 2012 by a polemical presidential decree which removed 780 km^2 from a number of conservation units of Pará in order to build hydroelectric reservoirs (MTV, 2011; WWF, 2012).

WHOSE PERCEPTION OF THE AMAZON COUNTS?

Amerindians and peasants of the Amazon are not alone in their struggle for land and for forest tenure reform. They are part of a larger global process in which local populations living in and around forests are pitted against both developmentalist interests who want to extract hardwood and clear land for commodity production and environmentalists with preservationist mentalities who until quite recently considered all humans to be a menace. In recent decades environmental policy has been decolonized so that local land use and environmental history are now understood on their own terms and native peoples and poor settlers are no longer seen as despoilers, poachers and thieves to be removed and turned into conservation refugees (cf. Adams & Mulligan, 2003; Dowie, 2009; Cawley, 2010; Fairhead & Leach, 1996; Jacoby, 2014). As a result between 1985 and 2000 over 200 million hectares of forest land in the world were legally transferred to or rights recognized for communities, mainly in Latin America, Asia, and Oceania (Larson et al., 2010).

Community-based forestry has thus become a third way which permits economic activity undertaken by local population living within conservation units. This tenure system represents a new form of agrarian reform because resource ownership and rights are redefined, land is appropriated and distributed and forestry rights are formalized. The system is based on common property resource management with characteristics and institutional innovations defined by local people for organizing and exercising their rights to use and manage forest areas in order to undertake logging and non-logging activities. As we saw in the case of the Tapajós National Forest, this meant giving priority to commercial sustainable logging.

In this paper local perceptions were shown to be fundamental for success because different forms of non-timber forestry favored by NGOs suffer a number of problems which restrict the composite income earned from them so that to date non-timber forestry has not contributed significantly to building viable green multi-functional livelihoods which could substitute frontier slash-and-burn agriculture and reduce carbon emissions in the Amazon in order to reduce global climatic change. Deforestation in the Amazon has been reduced from highs of 29,059 km^2 per year in 1995 and 27,772 km^2 in 2004 to lows of 4,571 km^2 in 2012 and 4,848 km^2 in 2014 (INPE, 2015) but in function of the IBAMA being more effective at enforcing environmental legislation and not due to the adoption of green alternative land use. Enforcement targets rancher and farmer settlers like those present near the Amazon National Park who are responsible for most of the deforestation which occurs in the Amazon but who are rarely contemplated for green solutions (with the exception of the failed attempt to grow native fruit for distant markets).

However, more than just including local perceptions is needed because the historical experience of agrarian reform shows that it is not enough to merely gain rights to land and local resources like INCRA has done since 1985 (Ghimire, 2001; Janvry, 1981; Mendras, 1970). Initial institutional development must go beyond internal community unity and negotiate across multiple scales in order to secure resource rights, supportive policy frameworks and appropriate technical assistance without giving rise to inequalities in benefit distribution. Ideally, this should evolve to an inter-communitarian scale through which local people connect with powerful decision makers, donors and the public in general without losing effectiveness and local legitimacy (Taylor & Cheng, 2012, pp. 110–111).

Much like the two cases of community forestry treated by Taylor and Cheng in Guatemala, the historic riverine peasants of the Tapajós National Forest and the Tapajós-Arapiuns Rubber Tapping Reserve were well suited for developing secondary-level associations, which provide a range of services, a forum for civic dialogue, learning, and advocacy of member interests. The associations of the National Forest eventually came together in a cooperative for channeling external assistance and direct provision of technical assistance and production coordination. By contrast, the peasants who settled inside the Amazon National Park are not originally from the Amazon and when threatened with eviction were able to organize themselves in order to resist and to negotiate a compromise concerning land tenure and public services but they did not go beyond that. The Tapajós National Forest and possibly the Tapajós-Arapiuns Rubber Tapper Reserve movements moved from building social capital to creating considerable human and economic capital while the settlers in the Amazon National Park never got past the social capital phase and consequently did not develop sustainable rural livelihoods.

However, it remains to be seen if the community-based forestry system developed in the Tapajós National Forest only was a success because of its close proximity to the important regional city of Santarém. Timber products can be transported long distances but political synergy in distant communities may never gel, in which case the experience of the Tapajós National Forest would merely be an interesting experiment but only a juxtaposed activity which does not contribute to sustainability in the Amazon at the regional level. The environmental agencies could make the difference. The agencies are now firmly committed to sustainable land use by local populations in a number of conservation units located across the Amazon. Over the last 15 years personnel has been renewed and young biologists, geographers, and sociologists entering the agencies have new perspectives that bridge the human-natural divide of Western conservationism of the past.

NOTES

1. Davidson et al. (2012) report a synthesis of recent research concerning deforestation and climatic change from the Large-Scale Biosphere Atmosphere Experiment in Amazonia. This project mobilized a large multi-disciplinary group of Brazilian and international scientists who carried out long-term experiments on the subject. Greater details are available at the project web page http://www.lbaeco.org/lbaeco

2. The best recent overview of this is provided in the Epilogue to the Second Edition of Hecht and Cockburn (2010) which ironically renders Chapters 1–9 reproduced from the First Edition (1990) to history.

3. The INCRA is also responsible for the modest program of land reform in Brazil but colonization projects in the Amazon hardly qualify as land reform unless one holds the questionable view that redistributing land from Amerindians and historic peasants to settlers from other regions of Brazil counts as land reform. During the military dictatorship from 1964 to 1985 comprehensive planned colonization projects were set up in the Amazon but after that date the INCRA concentrated on titling land occupied by squatters who arrived from other regions.

4. This system has been criticized for involving carbon colonialism because: (1) plantations can be considered forest to the detriment of local peoples and biodiversity, (2) carbon stocks become commodities while other non-commoditized elements of the local ecosystem are ignored, (3) the credits permit post-industrial countries to continue polluting because (until recently) they have the resources to finance projects elsewhere in the world, and (4) national sovereignty issues arise concerning who actually owns the carbon resources protected (Bumpus & Liverman, 2011, pp. 201, 217–281).

5. A 1965 law set the limit at 50% for farms located in forested zones of the Amazon and this was changed to 20% in 1998. Then in 2013 farms with land opened before 1998 when the limit was 50% were no longer required to reforest their land in order to comply with the 1998 20% limit. Savanna enclaves suffer less restrictions presently being required to preserve only 35% of the land. No doubt further change will occur as agribusiness operations move into the Amazon and pressure for easing the limits.

6. The classic definition of *caboclos* is "detribalized Amerindians, the offspring of European-Amerindian marriages, and descendants of early immigrants of varying ethnic and national origins" (Hiraoka, 1992, p. 135). This author also observed that despite being the majority of inhabitants in lowland riparian Amazonia they were at that time invisible in research and government policy which instead concentrated on tribal Amerindians or new settlers along roads. This occurred because the riverine population was dispersed, did not have representation in regional, national, and international forums and had non-exotic life-styles. When Hiraoka wrote the situation was changing and the book in which his chapter appeared was one of several books and articles focused on riverine peasants published after 1990: Chibnik (1994), Nugent (1993, 2002), and Padoch, Ayres, Pinedo-Vasquez, and Henderson (1999).

7. See Bicalho and Hoefle (2010) for a history of peasant political mobilization in the Central Amazon.

8. Moran (1982) and Hoefle (2003) provide detailed accounts of agroforestry and permaculture in the Amazon.

9. The homeland of the Munduruku is located about 400 km upriver and is an indigenous territory today. Murphy (1960) provided the classic ethnography of their intertwined history with the Portuguese colonial project.

10. See Nugent (2009) for a general critique of racial essentialism in the Amazon where a highly polygenic population is present which cannot be neatly encapsulated by a black/white/Indian system of classification. Further questioning the simplistic view of pristine tribal Amerindians, McSweeney and Jokisch (2007) showed that the greater majority of Amerindians in lowland Latin America live in cities and not in the remote indigenous territories set up for them.

11. The Tapajós-Arapiuns Rubber Tapping Reserve was first envisioned in the mold of the famous movement of the historic riverine peasants of Acre who resisted losing their traditional lands to private ranching interests. The Chico Mendes Rubber Tapping Reserve was the first of this kind of conservation unit which granted concession rights to land for developing latex extraction. Unfortunately the attempt was not an economic success and many of the peasants turned to cattle-raising which is banned from this kind of conservation unit. See Gomes, Perz, and Vadjunec (2012) and Rosendo (2010) for recent reviews concerning Acre.

12. This is a perennial problem for settlers which has been highlighted since the first studies of the colonization along new roads and continues to be a major impediment today along unpaved highways. For details see Aldrich, Walker, Simmons, Caldas, and Perz (2012), Bicalho and Hoefle (2008), Caldas et al. (2007), Foweraker (1981), Hecht and Cockburn (1990), Moran (1981), Schmink and Wood (1992), and Velho (1972).

ACKNOWLEDGMENT

Research funded by the Conselho Nacional de Desenvolvimento Científico e Tecnológico (CNPq – Brazil) and the Institut pour la Recherche de Développement (IRD – France). Fundação Carlos Chagas Filho de Amparo à Pesquisa do Estado do Rio de Janeiro.

REFERENCES

Adams, W. M., & Mulligan, M. (2003). Introduction. In W. M. Adams & M. Mulligan (Eds.), *Decolonizing nature: Strategies for conservation in a post-colonial era* (pp. 1–15). London: Earthscan.

Aldrich, S., Walker, R., Simmons, C., Caldas, M., & Perz, S. (2012). Contentious land change in the Amazon's arc of deforestation. *Annals of the Association of American Geographers*, *102*, 103–128.

Bicalho, A. M. S. M., & Hoefle, S. W. (2008). On the cutting edge of the Brazilian frontier: New and old agrarian questions in the south central Amazon. *Journal of Peasant Studies*, *35*(1), 1–38.

Bicalho, A. M. S. M., & Hoefle, S. W. (2010). Economic development, social identity and community empowerment in the central and western Amazon. *Geographical Research*, *48*, 281–296.

Boserup, E. (1965). *The conditions of agricultural growth*. Chicago, IL: Aldine.

Bumpus, A. G., & Liverman, D. M. (2011). Carbon colonialism? In R. Peet, P. Robbins, & M. Watts (Eds.), *Global political ecology* (pp. 203–224). London: Routledge.

Butler, R. A. (2010). *Threats from mankind*. Retrieved from www.mangabay.com

Caldas, M., Walker, R., Perz, S., Arima, E., Aldrich, S., & Simmons, C. (2007). Theorizing land cover and use change: The peasant economy of colonization in the Amazon Basin. *Annals of the Association of American Geographers*, *97*, 86–110.

Cawley, M. (2010). Adding value locally through integrated rural tourism. In G. Halseth, S. Markey, & D. Bruce (Eds.), *The next rural economies* (pp. 89–101). Wallingford, CT: CABI.

Chibnik, M. (1994). *Risky rivers: The economics and politics of floodplain farming in Amazonia*. Tuscon, AZ: University of Arizona Press.

Costa, S. M. G., Vieira, J. C., Lima, M. V. C. L., & Mileo, B. P. (2013). Territorialização e identidade indígena no baixo Tapajós no Estado do Pará. *IV Jornada Internacional de Políticas Públicas* (pp. 1–10). São Luiz: UFMA.

Davidson, E., Araújo, A., Artaxo, P., Balch, J., Brown, I., Bustamante, M., ... Wofsy, S. (2012). The Amazon basin in transition. *Nature*, *481*, 321–328.

Denevan, W. M. (1996). A bluff model of riverine settlement in prehistoric Amazonia. *Annals of the Association of American Geographers*, *86*(4), 654–681.

Dowie, M. (2009). *Conservation refugees: The hundred-year conflict between global conservation and native peoples*. Cambridge, MA: MIT Press.

Durham, W. H. (2008). The challenge ahead. In A. Stronza & W. H. Durham (Eds.), *Ecotourism and conservation in the Americas* (pp. 265–271). Wallingford, CT: CABI.

Eagles, P. F. J., & McCool, S. F. (2000). *Tourism in national parks and protected areas*. Wallingford, CT: CABI.

Fairhead, J., & Leach, M. (1996). *Misreading the African landscape*. Cambridge: Cambridge University Press.

FAS (Fundação Amazonas Sustentável). (2007). *Criação e significado da bolsa floresta*. Retrieved from www.fas-amazonas.org

FAS (Fundação Amazonas Sustentável). (2012). *8090 famílias cadastradas na Bolsa Floresta*. Retrieved from www.fas-amazonas.org

Foweraker, J. (1981). *The struggle for land*. Cambridge: Cambridge University Press.

Frake, C. O. (2014). How to be a tribe in the Southern Philippines during the advent of NGOs and the invention of the indigenous. *Human Organization*, *73*(3), 197–204.

GCP (Global Carbon Project). (2007). *GCP – Carbon budget*. Retrieved from www.globalcarbonproject.org/%20carbonbudget/07/index.htm

Ghimire, K. B. (2001). Land reform at the end of the twentieth century. In K. B. Ghimire (Ed.), *Land reform and peasant livelihoods* (pp. 1–25). London: ITDG.

Gomes, C. V. A., Perz, S. G., & Vadjunec, J. M. (2012). Convergence and contrasts in the adoption of cattle ranching. *Journal of Latin American Geography*, *11*(2), 99–120.

Hecht, S. B., & Cockburn, A. (1990/2010). *The fate of the forest: Developers, destroyers, and defenders of the Amazon* (2nd ed.). Chicago, IL: University of Chicago Press. (Original work published 1990).

Hiraoka, M. (1992). *Caboclo* and *ribereño* resource management in Amazonia: A review. In K. H. Redford & C. Padoch (Eds.), *Conservation of neotropical forests* (pp. 134–157). New York, NY: Columbia Univesity Press.

Hobsbawn, E. J. (1969). *Industry and empire*. London: Penguin.

Hoefle, S. W. (2003). Permaculture and regional rural sustainability in the Amazon. In A. M. Bicalho & S. W. Hoefle (Eds.), *The regional dimension and contemporary challenges to rural sustainability* (pp. 322–336). Rio de Janeiro: Laget-UFRJ/CSRS-IGU.

Hoefle, S. W. (2013). Santarém, cidade portal de fronteiras históricas do oeste do Pará. *Espaço Aberto*, *3*, 45–76.

Hoefle, S. W. (2014). Critical approaches to rural and eco-tourism: Sustainable development or green washing the Amazon? In A. M. S. M. Bicalho & L. Laurens (Eds.), *The changing face of the contemporary countryside* (pp. 197–220). Rio de Janeiro: Laget-UFRJ/CSRS-IGU.

Ianni, O. (1979). *Colonização e contra reforma agrária na Amazônia*. Petrópolis: Vozes.

IBAMA (Instituto Brasileiro do Meio Ambiente e dos Recursos Naturais Renováveis). (2005a). *Ecoturismo na Floresta Nacional do Tapajós*. Santarém.

IBAMA (Instituto Brasileiro do Meio Ambiente e dos Recursos Naturais Renováveis). (2005b). *Plano de Manejo: Floresta Nacional do Tapapós*. Brasília.

IBGE (Instituto Brasileiro de Geografia e Estatística). (2010). *@cidades*. Retrieved from http://www.cidades.ibge.gov.br

ICMBio (Instituto Chico Mendes de Conservação da Biodiversidade). (2014). *Floresta Nacional do Tapajós*. Retrieved from http://www.icmbio.gov.br/%20flonatapajos

ICMBio-Google Earth. (2014). *Mapa com limites: Parque Nacional da Amazônia*. Retrieved from www.icmbio.gov.br/portal/biodiversidade/unidades-de-conservacao/biomas-brasileiros%20/amazonia/unidades-de-conservacao-amazonia/1971-parna-da-amazonia.html

INPE (Instituto Nacional de Pesquisas Espaciais). (2015). *Projeto Prodes: Monitoramento da floresta amazônica brasileira por satélite*. Retrieved from www.obt.inpe.br/prodes/index

ISA (Instituto Socioambiental). (2014). *Território Indígena Bragança/Marituba e Território Indígena Munduruku-Taquara*. Retrieved from http://ti.socioambiental.org/pt-br/#!/pt-br/terras-indigenas%20/4178/4176

Jacoby, K. (2014). *Crimes against nature: Squatters, poachers, thieves* (2nd ed.). Berkeley, CA: University of California Press. (1st Edition published 2001).

Janvry, A. (1981). *The agrarian question and reformism in Latin America*. Baltimore, MD: Johns Hopkins Press.

Larson, A. M., Barry, D., & Dahal, G. R. (2010). Tenure change in the global South. In A. M. Larson, D. Barry, G. R. Dahal, & C. J. P. Colfer (Eds.), *Forests for people* (pp. 3–18). London: Earthscan.

Mathis, E. (1998). Garimpagem de ouro e valorização da Amazônia. Paper do NAEA 101. Belém: UFPA.

McSweeney, K., & Jokisch, B. (2007). Beyond rainforests: Urbanisation and emigration among lowland indigenous societies in Latin America. *Bulletin of Latin American Research*, *26*(3), 159–180.

Mendras, H. (1970). *Sociétés paysannes*. Paris: Librairie Armand Colin.
MICT/MMA (Ministério da Indústria, Comércio e Tourismo/Ministério do Meio Ambiente). (1994). *Diretrizes para uma política nacional de ecotourismo*. Brasília.
MMA (Ministério do Meio Ambiente). (2002). *Ecoturismo: visitar para conservar e desenvolver a Amazônia*. Brasília.
MMA (Ministério do Meio Ambiente). (2008). *Unidades de conservação apoiadas pelo programa áreas protegidas da Amazônia (map)*. Retrieved from http://www.mma.gov.br/estruturas/sbf_arpa2008/_publicacao/%20154_publicacao31032009094817.pdf
MMA (Ministério do Meio Ambiente). (2013a). *O sistema nacional de unidades de conservação da natureza*. Retrieved from www.mma.gov.br/estruturas/240
MMA (Ministério do Meio Ambiente). (2013b). *Programa para o desenvolvimento do ecoturismo na Amazônia Legal – PROECOTUR*. Retrieved from www.mma.gov.br/port/sca/proeco
Moisey, R. N. (2000). The economics of tourism in national parks and protected areas. In P. F. J. Eagles & S. F. McCool (Eds.), *Tourism in national parks and protected areas* (pp. 235–253). Wallingford, CT: CABI.
Molnar, A. (2011). Forest tenure reform in Indonesia. *Anthropology News*, *52*(9), 18.
Moran, E. F. (1981). *Developing the Amazon*. Bloomington, IN: University of Indiana Press.
Moran, E. F. (1982). Farming systems in the Amazon. *Culture & Agriculture*, *15*, 1–8.
MTV (Movimento Tapajós Vivo). (2011). *Governo vai reduzir unidades de conservação para construir hidroelétricas no Tapajós (PA)*. Retrieved from http://www.movimentotapajós%20vivo.com.br
Muirhead, B., & Almas, R. (2012). The evolution of Western agricultural policy since 1945. In R. Almas & H. Campell (Eds.), *Rethinking agricultural policy regimes* (pp. 23–49). Bingley, UK: Emerald Group Publishing Limited.
Murphy, R. (1960). *Headhunter's heritage: Social and economic change among the Mundurucu Indians*. Berkeley, CA: University of California Press.
Nugent, S. (1993). *Amazonian caboclo society: An essay on invisibility and peasant economy*. Oxford: Berg.
Nugent, S. (2002). Whither *o campesinato*? Historical peasantries of Brazilian Amazonia. *Journal of Peasant Studies*, *29*, 162–189.
Nugent, S. (2009). Indigenism and cultural authenticity in Brazilian Amazonia. *Goldsmith Anthropology Research Papers*, *15*, 1–18.
Oliveira, A. (1999). *A geografia das lutas no campo*. São Paulo: Contexto.
Padoch, C., Ayres, J. M., Pinedo-Vasquez, M., & Henderson, A. (Eds.). (1999). *Várzea: Diversity, development, and conservation of Amazonia's whitewater floodplains*. New York, NY: New York Botanical Gardens Press.
Peixoto, R. C., Arenz, K., & Figueiredo, K. (2012). O movimento indígena no baixo Tapajós: Etnogênese, território, estado e conflito. *Novos Cadernos NAEA*, *15*(2), 279–313.
Presidência da República. (2011). *Lei que cria a bolsa verde é sancionada pela Presidenta Dilma Rousseff*. Retrieved from http://blog.planalto.gov.br/lei-que-cria-o-bolsa-verde
Rabelo, E. L. (2006). *20 anos nos garimpos do Tapajós*. Santarém: Instituto Cultural Boanerges Sena.
Rennington, K., Renwick, A., & Burton, R. J. F. (2012). Western European approaches to and interpretations of multi-functional agriculture – and some implications of a possible neo-productivist turn. In R. Almas & H. Campell (Eds.), *Rethinking agricultural policy regimes* (pp. 73–97). Bingley, UK: Emerald Group Publishing Limited.

Rosendo, S. (2010). *Evolving participatory institutions for tropical forest management: Adpatations and change in the context of extractive reserves in Brazilian Amazonia.* CSERGE Working Papers EDM 03-11, 1–22.

Schmink, M., & Wood, C. H. (1992). *Contested frontiers in Amazonia.* New York, NY: Columbia University Press.

Simmons, I. G. (1989). *Changing the face of the earth.* London: Routledge.

Stronza, A. (2008). The bold agenda of ecotourism. In A. Stronza & W. H. Durham (Eds.), *Ecotourism and conservation in the Americas* (pp. 3–17). Wallingford, CT: CABI.

Taylor, P. L., & Cheng, A. S. (2012). Environmental governance as embedded process: Managing change in two community-based forestry organizations. *Human Organization, 71*(1), 110–122.

Torres, R. M., & Momsen, J. H. (2011). Introduction. In R. M. Torres & J. H. Momsen (Eds.), *Tourism and agriculture* (pp. 1–10). Milton Park: Routledge.

Vaz Filho, F. A. (2010). *A emergência étnica de povos indígenas no baixo Rio Tapajós, Amazônia.* Doutorate thesis, Universidade Federal da Bahia.

Vaz Filho, F. A. (2011). Povos indígenas no baixo Tapajós querem reconhecimento. In C. A. Ricardo & F. Ricardo (Eds.), *Povos indígenas do Brasil 2006/2010* (pp. 423–426). São Paulo: ISA.

Velho, O. G. (1972). *Frentes de expansão e estrutura agrária.* Rio de Janeiro: Zahar.

Viana, V. M. (2008). Bolsa Floresta. *Estudos Avançados, 22*(64), 143–153.

Viana, V. M. (2011). *Bolsa floresta e bolsa verde: semelhanças, diferenças e desafios.* Retrieved from http://envolverde.com.br/ambiente/legislacao-ambiente/bolsa-floresta-e-bolsa-verde

Wilson, G. (2010). Multi-functional 'quality' and rural community resilience. *Transactions of the Institute of British Geographers, 35*(4), 364–381.

Woods, M. (2011). *Rural.* Milton Park: Routledge.

WWF (World Wifelife Fund). (2012). *Tapajós: o rio de vez para construção de hidroelétricas.* Retrieved from www.wwf.org.br

GLOBAL WARMING OR CASH ECONOMY? DISCOURSES OF CLIMATE CHANGE AND FOOD IN EASTERN HIGHLANDS PROVINCE, PAPUA NEW GUINEA

Olivia Barnett-Naghshineh

ABSTRACT

Purpose – *This paper describes the different ways in which people in the highlands of Papua New Guinea are talking about climate change. It demonstrates that people locate themselves in this process of change in terms of food production and exchange, and that some of the changes being witnessed are also related to the impacts of a growing cash economy on social relations.*

Methodology/approach – *This ethnography involved 12 months fieldwork including participant observation and interviews.*

Research limitations/implications – *This is a qualitative study that recognises the perspective of local people for understanding culturally mediated experiences of climate change. However, data regarding*

Climate Change, Culture, and Economics: Anthropological Investigations
Research in Economic Anthropology, Volume 35, 107–133

ISSN: 0190-1281/doi:10.1108/S0190-128120150000035005

rainfall and temperatures over time would be a useful addition for thinking about the extent to which the climate has in fact changed in recent years.

Practical implications – *The implications of this paper are that the predictions made in 1990 about increases in production as a result of climate change are apparently coming true, with benefits for some food and coffee producers. But that there are complex social processes occurring at the same time as climate change that mean people's ability to adapt is dependent on other social conditions. Maintaining ecologically sustainable methods of production and local cultural practices may enable more resilience to the impacts of climate change.*

Originality/value – *The experiences of people living in the Eastern Highlands and the ways in which people use the discourse of climate change are yet to be acknowledged in policy circles or socio-cultural anthropology literature. This paper presents a partial account of how people in Papua New Guinea are experiencing and talking about change.*

Keywords: Climate change; food production; exchange; Papua New Guinea; discourse; environmentalism

INTRODUCTION

When I commented on the intense heat in Goroka (capital of the Eastern Highlands Province (EHP) Papua New Guinea) to the young driver of a pubic town bus, he remarked,[1]

> Yes, it is very hot in Goroka now, it wasn't like this before but now we have climate change. Before we would always wear coats and jumpers, now you can go out in just a t-shirt! Plus we can grow mangoes now! And they are sweet and juicy ones, much nicer than the mangoes from the coastal areas!

In this paper, I show that people in Goroka are experiencing climate change, both in terms of unpredictable seasons and increased temperatures which affects what types of food they can grow and where, but also as a discourse that is influencing people's perceptions of change. I demonstrate that people in Goroka understand climate change in relation to food production and exchange because this is such a key part of their lives, both selling food as a commodity and gifting food for maintaining social relations (Fajans, 1988; Meigs, 1984; Curry, 1999; Meggit, 1974; Maclean, 1989; Kahn & Sexton, 1988). The discourse of climate change not only frames what people are seeing as different in their local food landscapes but also validates changes that are occurring. But 'climate change' as a discourse comes from Western scientific contexts and can obscure, obfuscate and confuse local understandings of when change is ecological, environmental or socio-economic. This matters for how people understand their place in the world and how they see themselves related to the changes that are occurring.

I proceed by first addressing why anthropology is well placed to consider the interconnections between global warming and climate change discourse in local contexts. Then I locate PNG and the highlands within broader discussions of climate change, before providing an ecological and historical overview of where I carried out my fieldwork. Relating this to food production over time, I provide an ethnographic analysis of how people conceptualise and narrate climate change in Goroka today with a particular focus on food production and exchange, through case studies of three different crops: winged bean, watermelon and african yam. In the conclusion, I draw together the implications of these accounts for thinking about how people in Goroka are experiencing and understanding different kinds of change.

THE GLOBAL AND THE LOCAL THROUGH AN ANTHROPOLOGICAL LENS

Climate change is occurring worldwide (Jacka, 2009), but how and with what effect depends on the social and geographical context (Fiske et al., 2014; Orlove, Lazrus, Hovelsrud, & Giannini, 2014; Vásquez-León et al., 2003). Lahsen (2004) argues that the global discourse of climate change collectivises the problem of the earth's warming and increasing weather unpredictability, making it appear as though the global North and South are in a mutual struggle. This creates 'a cognitive and normative understanding that transcends local, parochial interests' and makes climate change a global moral issue (Lahsen, 2004, p. 153). Such universal discourses can hide contradictions as well as the socio-political realities of particular contexts, for example, rural communities that rely on agricultural production for survival are assumed to be hardest hit by climate change (Lahsen, 2004).

Furthermore, the geographic areas researched are often pre-defined by policy frameworks and institutions assigned to climate change, which leaves some parts of the world unexamined (Orlove et al., 2014). This approach also puts hard data above that of local experiences (Strauss & Orlove, 2003).

Climatologists and geographers use hard data over large regions or long periods of time to understand how climate varies temporally and spatially.[2] This privileges external data such as meteorological records over more interpretive perspectives of local actors (Jacka, 2009). Meanwhile in the absence of long-term climatological records, local observations can provide information where gaps in data exist (Fiske et al., 2014; Jacka, 2009). Thinking about the future of climate change and its impacts on different societies requires an examination of how impacts are culturally mediated and differ between ecological zones. Furthermore, 'climate change' has become a global discourse that people apply to their own contexts.

Tsing (2005) shows with the example of environmentalism in Indonesia, how global interconnections, both as discourse and practice, are sometimes messy and confused. When an external discourse is defined by 'experts' and introduced by government agencies and media, different people can then apply it according to their own interpretations and observations, creating a new unique perspective. Using the same language does not imply they are necessarily talking about the same thing (Tsing, 2005). I see climate change as one such discourse, in no way denying the occurrence of global warming, but recognising that the discussion of this phenomenon in global policy circles and environmentalist networks has influence at other levels. Rudiak-Gold (2011) calls for 'reception studies', using anthropological methods, to investigate how communities receive, interpret and adopt the discourse of climate change. Bearing this in mind, I show what 'climate change' means for people in Goroka and how people use this term for their observations of change.

In the Eastern Highlands, people often discuss climate change in relation to food production, something that is morphing rapidly into a means of creating capital flow and becoming less about subsistence or gift exchange. Goroka, its fresh food market, and the many food producers surrounding it provide what Tsing refers to as a 'frontier culture', those most recently drawn into the world system and increasingly affected by Western capitalist expansion (Tsing, 2005). Additionally, PNG has been subject to development agendas such as 'Rural Development' under Australian mandate and despite political independence in 1975. This has focused on commercial agriculture for economic growth and tried to integrate smallholder farmers into the country's economy, making Goroka a site of complex interconnectedness between global trends, national development and evolving local cultural practice (Benediktsson, 2002; Escobar, 1995; Fleming & Hardaker, 2010).

CONTEXTUALISING PNG AND THE HIGHLANDS

Due to concerns about rising sea levels in the Pacific, PNG, along with many of the Small Island states, has been the focus of international organisations and legal climate change frameworks (Barnett and Campbell, 2010). Lipset (2013) has shown how rising sea levels have disastrous consequences for people who are dependent on the land in low-lying areas of the country. In 2009, the UN Office for Humanitarian Affairs reported people from the Carteret Islands, off the coast of Bougainville, fleeing their homes as the islands went under water and the community became the first refugees of current global warming (Monbiot, 2009). Furthermore, in 2013, a study by the Asian Development Bank identified the economy of PNG as likely to be worst hit by climate change in the Pacific. This is partly because of land depletion with rising sea levels but also predicted failures in sweet potato crops (a major agricultural product) (Sullivan, 1990, 1991). The report warns of a potential loss of 12.7% of GDP by 2100 relating to the environmental changes that will take place. 'Along with a temperature spike of nearly 3°C, PNG is likely to see more frequent and heavy rains, with the sea level potentially rising by up to 1.45 meters by 2100' (Asian Development Bank, 2013).

However, the highlands (see Fig. 1) are rarely mentioned in the media or by development agencies as an area of concern in relation to climate change, to the extent that the National Office for Climate Change and Development does not carry out research in the highlands nor has any projects there.[3] The highlands are noted, both by researchers and locals, to be quite different from the rest of PNG. In this case, the major difference is that people are not directly at risk from rising sea levels. It is, however, a region that has undergone significant change in a relatively short period of time (mainly since the 1950s), compared to many other parts of the country

The highlands are in the centre of the island, 1,200 m above mean sea level, and are possibly one of the earliest sites of horticulture in the world with evidence that people were cultivating taro 10,000 years ago and forest

Fig. 1. A Physical Map of Papua New Guinea. *Source*: Copyright Anthropology, Social Sciences, University of Auckland, 2015.

clearance dating back to 30,000 years ago (Bourke & Harwood, 2009; Denham et al., 2003; Sutton et al., 2009).

Since then there has been a long history of food cultivation and new foods entering the region. For example, after Europeans colonised the Americas, an array of crops was introduced to New Guinea via Southeast Asia (such as corn and passion fruit) (Bourke & Harwood, 2009; Sillitoe, 2010). People in the highlands are known for their ability to promulgate new crops and be creative in the process, since people have been gardening here for such a long time (Bourke & Harwood, 2009). Food is a major component of social and cultural life in PNG (Fajans, 1988; Maclean, 1988; Read, 1965; Kahn & Sexton, 1988). For example, sweet potato was introduced to the highlands about 300 years ago and has been a staple since, shaping the agricultural system and importantly allowing for the breeding and consumption of pigs – a major wealth item for ceremonial exchanges and compensations (Bashkow, 2006; Bourke & Harwood, 2009; Hide, 1984; Finney, 1969; Keil, 1974; Kyakas & Wiessner, 1992;

Rappaport, 1968; Sillitoe, 2001; Sexton, 1986; Strathern, 1988; Strathern & Stewart, 1999).

FOOD PRODUCTION IN GOROKA

My research in Goroka was mainly with people who grow food and sell it in the main market. Today Goroka is the second largest town in the Highlands region and has a population of about 40,000 (Bourke & Harwood, 2009).[4] EHP was recorded as 91% rural and 9% urban according to the national census carried out in the year 2000 (Bourke & Harwood, 2009). Goroka town sits in a river valley that carves out the middle of the EHP and is encircled by mountains. Gardens can be seen all over the landscape, neat rectangular areas where the land has been cleared; many of these gardens supply food to the market. People live in the river valley and in the mountains, sometimes at altitudes of 2,000 m above sea level. Whilst there are generally understood to be two seasons, a wet season and a dry season, local geography means some parts of the Eastern Highlands can be wet while other parts can be very dry (Brouwer et al., 1998). This also affects the soil types and what kinds of food can be grown where. The Eastern Highlands was the first part of the highlands to be entered by gold explorers in 1933, the initial 'first contact' being made by the Australian Leahy brothers (Munster, 1986; Sexton, 1986).

The Goroka market was set up in 1952 by the Australian authorities, who had been granted mandate of New Guinea after the Second World War, in agreement with local leaders including the adopted father of today's Provincial Governor Julie Soso. This was the beginning of locally produced food being exchanged for money in Goroka. Elders told me that prior to this, they took their sweet potato and other crops thought desirable by expatriates (such as tree fruits), to the original colonial station (now the University of Goroka) where they could exchange their food with Europeans for salt. They would take empty pieces of bamboo to be filled up with salt by the expatriates and carry it back to the village. The expatriate community existed mainly due to the 1950s expanse of European plantation coffee and setting up of mission stations:

> In 1950 there was a European population of about fifteen, by 1966 the town had 4816 inhabitants, including 1000 Europeans ... Goroka was a thriving European commercial and administrative outpost in the New Guinea highlands. (Finney, 1969, p. 7)

Missionaries and colonial authorities encouraged local people to cultivate crops for expatriate consumption and also because they deemed a varied diet would be beneficial for 'the natives'. Tomatoes, potatoes and carrots were introduced and people living close to Goroka were taught how to propagate them for sale to expatriates living in town. The emerging local elite also purchased these food crops, those who were educated and joined the public service or worked for local businesses (Errington & Gewertz, 1999). EHP has a relatively short (around 60 years, nearly three generations) history of producing food as a commodity, with money itself only being introduced in 1947 (Sexton, 1986; Waiko, 1993). Some people have developed highly successful businesses from food production, while others, perhaps with less land, labour or substandard soil fertility, grow barely enough to make ends meet through local markets. There is a struggle over resources, mainly land (Brookfield & Brown, 1963). Furthermore, people use fertilisers and pesticides to grow foods that are not 'local' but can reach good prices in the market such as broccoli, which can and does lead to a decline in soil fertility if over used.

Many of those selling in the Goroka main market are what Epstein (1982) defines as 'producer–sellers', selling surplus food from their gardens. These were predominantly women, though I also spoke to men selling in the market as well. Today, it is mostly women who take food to market, and it is largely women who are responsible for producing the food too, whilst men tend to concentrate on cash crops, principally coffee (Benediktsson, 2002; Brookfield & Brown, 1963; Sexton, 1986; West, 2012).

The extent to which people can sell food in the market partly depends on their proximity to town and the quality of their road networks. Those communities that are well connected to urban road networks are likely to produce a greater share of their crops for the market than those who have to pay high bus fares and walk long distances. I carried out research with communities of varied proximities. However, a distinctive characteristic of the Eastern Highlands is that people will grow food for the market, however, far it may be to travel and walk miles of mountainous tracks in order to get their crops to market. A vast majority of food producers around Goroka harvest at least some of their food for sale in the market, if not most of it. This is a dramatic change that has come about in the last 50 years.

Since the 1980s, there has been a growth in urbanisation and an increased desire for cars, mobile phones and other commodities that represent modernity and more 'western' lifestyles (Errington & Gewertz, 1999;

Foster, 2008; Akins & Robbins, 1999). To fund their children's education in hope that they will seek out paid jobs, many horticulturalists in the Eastern Highlands focus their production increasingly on crops that can be sold for decent money in the market (often those introduced since colonisation such as broccoli and cabbage) to provide an income that can sustain their families' material needs and desires.

Such changes in lifestyle seem to have occurred mainly, but not exclusively, in the periphery of Goroka town where people replace their bush-material houses (of grasses and bamboo) with semi-permanent or permanent houses using materials such as aluminium roofs and cheap fibreboard walls. Besides providing more durable shelter, this transformation is symbolic to them of progress and affluence. One market seller was able to pay for her family to have a permanent house having earned and saved her money from fresh produce marketing. She explained how this meant she need not worry about the house collapsing or disintegrating when she and her husband grow elderly. Continuous rebuilding of houses was a normal process prior to European contact and economic development, and required skills of men in the community which, I was told frequently by people in Goroka, are rapidly being lost.

The pull of modern and urban life has also led to significant out-migration from rural villages. Thus as Goroka town grows, villages in the exterior lose their young men and women to better schools and work opportunities in town or, as recent census data has shown, to the peripheries of town (PNG Census, 2011). These changes seemed more pronounced today than previous years for many of the older people I spoke with. Younger generations no longer spend the majority of their time in the garden as their parents did, but attend school instead – one symptom of a government provided 'free education'[5] and the realisation that education is a pathway to paid work. The subsistence lifestyle that had long provided the foundation for cultural life in the Eastern Highlands has changed.

Many highlanders recognise the widespread desire for money and material goods as a relatively recent phenomenon (the last 30 years or so) with which their parents' generation were not so concerned. For example, as a group of food producers from the village of Kasena in Daulo district loaded up a truck with carrots opposite Goroka market, to be sold in Lae city, they told me that their parents had not done the same they were doing, that is, producing cash crops and moving them to places where they will get higher returns.

LOCAL PERSPECTIVES ON CLIMATE CHANGE

Predictions were made in the early 1990s about how climate change would impact upon the highlands; Hughes and Sullivan (1990) stated that this area would become more agriculturally productive, relieving the pressure of a growing population, as more land would become fertile for subsistence and coffee production. The accounts that I have collected from producer–sellers, commercial farmers, coffee industry workers, agricultural extension workers and others around Goroka mostly align with Hughes' prediction – that areas previously too cold have become productive. But this is just one aspect of how people are being affected by climate change.

When I walked through people's gardens and talked to them about the crops they were growing or interviewed them about their produce in the market, they would often make reference to 'climate change'. Agricultural extension workers[6] would mention it as they told me about their work, explaining how they teach people to deal with the impacts of climate change, such as droughts and an increase in pests. They were encouraging the planting of new and more resilient food crops or the reorganisation of gardens so that water could be accessed if and when the rains failed to come. Some of this work had begun with the devastating impacts of the 1997 drought in PNG brought on by El Nino (Bourke & Harwood, 2009). Farmers in EHP were relatively unscathed compared to other parts of the country – but one eighth of PNG's population was without rain, food and water for one of the longest periods in 100 years (Bourke & Harwood, 2009; Ellis, 2003; Eves, 2000). It was clear that people in Goroka were convinced that climate change was happening; however, there were varied understandings of what this meant.

One active producer–seller, Mama Esa, told me on our first meeting in her village (on a quiet Sunday surrounded by people passing time by playing cards) that something called climate change[7] is affecting them. When I asked what she thought was bringing on these changes, she said, 'the sun's system is coming closer so things are getting warmer'. A young man overheard our conversation and corrected her, 'No, there is a hole in the ozone layer due to all the factories and cars pumping out smoke, and now the North Pole and Arctic are melting as the Earth's getting warmer. And sea levels are rising'. Mama Esa responded, 'I don't know why it is happening, now is just a time of change. I see all of these things as part of that change'. The young man's ability to talk about the ozone layer and what is going on elsewhere in the world likely indicates a formal education which Mama Esa never had, as well as access to radio and mobile phones.

For Mama Esa, and many other producer–sellers I spoke with, climate change meant an inconsistency in seasons, 'em go longlong olgeta' – it's gone completely crazy, she said. The beginning of the wet season was once clear; the rains would come suddenly and consistently at a similar time (around October). Now, it is no longer clear when the dry season or the rainy season is coming. During an interview at the University of Goroka's farm, Gure, an Agricultural Officer who has been an advisor to farmers in the Eastern Highlands for over 20 years, explained:

> This has a big impact when people want to plan their farming. During the dry season they used to prepare the land and during the wet season they would do planting and harvesting. But that has really been thrown into chaos. People don't know when they should start planting, and when they should harvest. Because of these sporadic rainy seasons, sporadic dry seasons, people are confused.

Climate change data has been collected for nearby Indonesia and northern parts of Australia and suggests similar trends (Bourke & Harwood, 2009).

For many, however, the biggest signal of change in EHP has been coffee production. People commented on the increasingly unpredictable coffee seasons with some trees bearing fruit much earlier than usual. May and June were known as the coffee months but now harvesting can happen as early as February, sometimes with a second round of picking at Christmas. I spoke with a long-term Australian employee of the Highlands Coffee processing factory in Goroka who said:

> Without a doubt something has changed, we now receive coffee beans all year round, rather than a big haul in June we are now processing bits of coffee more often.

As Hughes and Sullivan (1990) predicted, coffee now grows at altitudes that were previously considered too cold (1,800 m+). John, a local advisor to small scale farmers and growers cooperative, told me that smallholder coffee producers are in a good position to adapt to this change in temperature as they are able to care for their trees more closely than the plantations. He also said, however, that the increase in temperatures has meant a major problem with pests and more people turning to chemical pesticides to try and control them.

Mama Esa's observations of what changes were occurring were mainly regarding the location of food. As the eloquent daughter of a local clan leader who knew the history of the area, she was aware of which crops would normally grow in her village and which crops were known for growing closer to the valley basin. The way in which people in EHP understand the landscape is often related to what grows there. As Ives (2014) shows with Rooibos in South Africa, the plants that grow in a particular area partly construct 'place' for the people that are dependent and engaged with their ecological context. In the highlands, people of different language groups are associated with different ecological zones but form relationships across these zones through the exchange of food. For example, the Eastern Highlands has the Karuka-Marita[8] exchange – two types of pandanus that grow at different altitudes (Keil, 1974). A village or community in the cooler 'bush' areas would give large amounts of the karuka nut, which grows high up in trees only found in coolers parts of the highlands, to a community in a 'kunai' area (Tok Pisin for grassland), where the marita would grow as it required drier and hotter areas (Keil, 1974). When the season of marita arrived, the communities living in kunai areas would reciprocate with the marita plant. Although the seasons for karuka and marita slightly overlap, the exchange is not instantaneous and would normally be repaid within some weeks or months.

For Mama Esa, a major change has been in the production and exchange of particular crops, such as bananas and winged bean. Some types of bananas that could only grow in the lower altitude 'kunai' areas now grow in their local area – a much cooler and wetter area higher up in the mountains, which she referred to as 'bush place'.

Mama Esa explained that climate change has meant that the exchange of foods between the two locations of bush and kunai has stopped. Previously, she said, their relatives who had married into these villages or relatives of the mother's side who lived in these different ecological zones would bring them certain foods that they themselves would not eat otherwise. To reciprocate, the people of the mountain community would take down some of their food produce to the lower altitude kunai villages at another time.

Much of the food that would have been exchanged between these ecological zones is now growing in Mama Esa's village she said, so they have no need to exchange. Local agricultural extension workers and food producers both explain this as because of increasing temperatures in previously cooler areas. Thus, mobility of crops to different ecological zones is affecting the way people relate to their environment and each other, leading to a potential breakdown in social relationships, or at least a change in how people enact them (cf. South African Rooibos producers affected by climate change, Ives, 2014).

However, these kinds of exchanges, referred to as 'barter' in Tok Pisin, are likely to have subsided for an array of reasons. I discussed this with another woman in the market Legeniso, from Bena district and roughly about 40 years of age, who recalled how when she was at primary school, her village received a large basket of karuka nuts from a family that had recently become their in-laws. She remembers her uncles and aunts some months later harvesting lots of marita nut, which were then transported by car to the in-laws living in a bush area at a village named Kotoni. She explained that this exchange no longer occurs in her community because people concentrate on the production of food for market instead. Considering how full the market is of karuka nut during its season in January–March, this seems quite possible. However, Legeniso seemed to imply it is not just people selling their karuka at market, but that their food production in general is geared towards sale rather than sharing or exchanging with others. 'Before people didn't know about money too much' she reasoned, 'There wasn't many ways to use money. Now is money time'.

Although both Mama Esa and Legeniso both noticed change in patterns of food exchanges across ecological zones, Esa has used the term climate change to explain why this has occurred (as did local agricultural extension worker), while Legeniso saw the increased value given to money as a major factor. Legeniso was not alone. Another woman, who runs the Kafe Women's Association, a local NGO based in Goroka, shared a similar sentiment regarding the growing importance of money and the impacts upon social exchange of food. Her organisation now provides bags of rice to people who suffer poor harvests when previously these people would have received food from nearby villagers, who upon receiving word that people further down (or up) the mountain were going hungry would take surplus food to them. Today, she explained, many people have left the villages and are living in town, so there are less people working in the gardens that can contribute like this. For some, climate change explains such changes in the Eastern Highlands, while others recognise the role of a growing cash economy and urbanisation in shaping the choices people are making.

CASE STUDY I: WINGED BEANS AND SOCIAL CHANGE

There have been major shifts in the kinds of food people exchange in the highlands to recognise their social relations, a common Melanesian practice. I employ the winged bean (*Psophocarpus tetragonolobus*), locally known as 'asbin', as a case study of a previously highly valued food that is no longer being produced on the same scale. This illuminates how people are relating to some crops differently today, which in turn affects how people relate to one another, and the way in which climate change is used to explain this.

Winged bean is considered a very special crop in the Eastern Highlands, Western Highlands and parts of the neighbouring Wahgi valley, with a high socio-cultural value that meant it was a significant part of collective social exchanges before money entered people's ceremonial and material lives. In part, the value is derived from the crop itself as multiple parts of the plant are edible and its small tubers keep people satiated for a lengthy time period. Unlike other food crops where you are only able to consume the tubers or only the leaves, with the winged bean you can consume multiple parts of the plant as it grows; the flowers, beans, leaves and tubers. The tubers are cooked in the earth oven and given to people at significant moments and occasions to show thanks and appreciation.

Winged bean grows once a year with a growth cycle of about nine months in Goroka. People in Goroka emphasise the careful attention that this crop requires, more so than other ceremonial crops such as taro or yams. Owing to its distinct seasonality, winged bean was planted and thus harvested at the same time, especially for pre-organised ceremonial exchanges. Collected in village piles, the tubers were distributed to the people selected from other communities by leaders of the village. This made winged bean an important social crop for conveying certain messages and maintaining various kinds of relationships. It also contributed to the reputation of the producers and people giving the crop for being capable and talented.

As we sat in her village, Mama Esa gave me an example of when someone may use winged bean to show appreciation and thanks:

> If a woman is pregnant and can no longer move around freely, and someone brings her some firewood, somebody else brings her food from their garden and someone else brings her water, when she has had the baby and can move around again, she will go to her garden and plant *asbin*. About 9 months or a year later she will harvest the *asbin* [as tubers] and give them to those people who helped her. She will say thanks by giving these, and the people will be very happy with that.

She also told me that people are no longer planting winged bean on the same scale or with the same frequency. Gure elaborated, 'People are finding it difficult to know when to act out each particular task required for it to grow well'. The location of the sun and the moon in the sky is no longer corresponding to weather patterns and seasons. Due to seasonal variability, winged bean now grows at different times and speeds. Mama Esa said only a few women still plant it and they now do so on their own time schedules, no longer corresponding with others. This concurs with what many women who sold winged bean in the market told me. The winged bean tubers are still given to people to show thanks. But often only some are given while the best ones are sold. In places close to Goroka town, it is rare for them to be given as collective payments between communities.[9]

Both Esa and Gure explained that the decline in winged bean as a ceremonial crop was due to climate change, the unpredictable seasons and a disjuncture between seasons and the solar and lunar cycles. But my research suggests that this is just one strand of a more complex set of reasons. Winged bean has declined in its socio-cultural value, while money has become more valuable in the realm of collective ceremonial exchange; this incentivises growing crops that are more lucrative and less hard work.

The growing of winged bean is highly labour intensive and many women told me they knew how to grow it, but that it had become inconvenient to do so (Barnett-Naghshineh, 2014). For example, Sara, who was working as an elementary school teacher, told me about an event where she was going to thank a friend for supporting her daughter through school and reciprocate the woman's contributions to her daughter's initiation event. She said normally she could have used winged bean to show her appreciation; however, she had not planted it in years and instead was using money, from her wages to buy gifts (such as mattresses, pots and pans) instead.

Sara explained that she stopped growing winged beans when her husband had left her for another woman and she became a single mother. She that a man is needed to carry out the heavy labour involved in the preparation and cooking of this crop before exchange. Also, cash crops such as broccoli that are considered less hard work by many producer–sellers and shorter growth cycles appeal more. This kind of production brings in a constant income and allows people to use their monetary income for multiple ends, rather than investing their time and energy into a crop that is harder work. As Igoe et al. argue,

> A commodity is valued in relation to its ability to provide returns on investment or to generate additional capital value, while other types of value based on socio-cultural factors ... either fade into the background or disappear altogether. (Igoe, Neves, & Brockington, 2010)

As commodity production becomes more compelling, the winged bean declines and disappears from the local economy. This process is discursively subsumed under the rubric of change in seasons. Some people choose to use the term climate change, while others explain socio-economic changes. Sometimes this might be the difference between a generalised explanation and a more personally specific one. But using climate change is also a way of explaining the cause as something that is not local and thus impossible to affect.

On an aside, the overall decrease in large-scale exchanges of local food crops between communities that many in Goroka observed is potentially detrimental for an unstable climate future. While people would have previously given food to their neighbours during times of drought, I was told that people generally now think about themselves and their immediate family only. If climate change means severe droughts or severe rainfall, destroying people's gardens in the process, the networks of social relations, outside of ceremonial contexts, which once ensured food was shared, are disappearing.

CASE STUDY II: WATERMELONS, OPPORTUNITY AND INNOVATION

There are some changes in the Eastern Highlands that quite clearly relate to the local climate. An overall temperature increase was undisputed among the many people I spoke with, as well as two further observations: firstly, that there are crops known for being grown in coastal parts of PNG that are now bearing fruit in the highlands. And secondly, that trees and plants bear fruits in locations where it was not previously possible. I use watermelon to illustrate these shifts, which offers an explanation for why people in Goroka have so willingly taken on the term climate change to explain their observations.

During my fieldwork, only one woman was selling watermelons in the market that she herself grew, Julie from Bena district. She claimed to have pioneered growing pineapples in the 1990s (although other sources have said pineapples were introduced by missionaries in the 1970s). She told me that by the 2000s she was fed up with pineapples as there was too much competition from other sellers in the market. Pineapples are a desirable crop as they are a lucrative cash crop but do not require the same level of care as coffee, plus women are able to access capital from pineapples more easily than coffee.

Julie decided to try growing watermelons instead after one had mysteriously come up in her garden. She believes some visitors from the hot and humid town of Madang must have brought some of the fruit with them and then discarded the seeds on the ground. Once she knew watermelons could grow in her garden, she bought the seeds from a hardware store in Goroka town and planted them. For six years not one watermelon grew, she said. But then one year, her garden started to fill with these luscious fruits. She explained as we chatted in the market, 'Before Bena was a cold place, now it is hot – it's because of climate change!'

Julie told me watermelons are now the best thing to sell in the market; in one day she can make up to 600 kina (USD240), whereas pineapples may be around 200 kina (USD81) on a good day. On very successful days, she has even taken 1,000 kina (USD405) in one day – this is about eight week's wages for someone working in a local supermarket. Julie and her husband have been able to use the proceeds from her marketing to save up and purchase a Toyota Hilux (a popular consumer good), she told me proudly.

Julie is not the only one who benefitted from this new opportunity to grow watermelons. She explains how people come and buy her watermelons, then take them elsewhere in town and cut them into slices to sell as snacks, cutting smaller slices than she does but asking the same price. These types of sellers are known in Goroka as 'black market' sellers (Busse, 2012) and tend to purchase fresh produce from the market, then decrease the volume per unit and resell in other parts of town (or alternatively maintain the unit size and double the price). 'Black market people will come and buy five kina or ten kina watermelons and then go and cut them up into small pieces which they then sell for fifty toea or one kina', said Julie. The reselling by Goroka's black market sellers shows that not only do producers gain from growing novel crops in the highlands, but others also find opportunity in new commodities. The resellers are often urban settlers who do not have land to grow their own food and must find ways to generate an income to buy food instead.

Julie smiled when she told me the story of her watermelon as we sat under an umbrella, protecting ourselves from the blistering sun. It was early on in my research when she beckoned me over in the market saying, 'you are always talking to people in the market why don't you come and talk to me?' This was one of those serendipitous moments in fieldwork where people and their stories present themselves unexpectedly. She told me that her watermelons finally began to grow because of climate change. She explained that her part of the province is much hotter now (although it has always been warm, 'Bena bena' said to mean 'fire fire' because of common bush fires). In various situations, I heard Julie referred to when people discussed climate change – the woman that grows watermelons in Goroka. Just before I left Goroka I bumped into her again and asked if she knew whether anyone else was growing watermelons since we had first met nine months previously. She said this year she suspected there would be two or three others as she had seen people in her village trying to grow them.

CASE STUDY III: AFRICAN YAMS, ADAPTATION AND THE STATE

Another crop to be cultivated in the highlands recently is the white yam (*Dioscorea rotundata*), known locally as the african yam, which was introduced to Papua New Guinea in 1986 at the Saramandi Research Station in the East Sepik (Risimeri, Gendua, & Maima, 2001). It originates from West Africa where it is widely cultivated and has special socio-cultural status. The seeds were transferred to PNG from the 'Tropical Root Crop Improvement Program' in Nigeria. It is a particularly useful crop to grow as it can be easily multiplied through the 'mini-set' technique where one tuber is cut into smaller pieces and replanted. Seven months after planting they are ready for harvest. It was introduced to farmers in PNG following the 1997 drought. So that they are more resilient to droughts, people are being encouraged by the State to produce the african yam. This encouragement has resulted in the african yam becoming a popular crop that is used both in ceremonial exchanges and as a cash crop, sometimes making rare appearances in the Goroka food market. The high yield of the african yam makes it an attractive crop where there are land and population pressures (Risimeri et al., 2001). For example, I had friends who were growing a few african yams in their back gardens in Goroka town, as well as in gardens further out of town.

Vice-president of the Eastern Highlands Women in Agriculture (WIA) group at the time of my research, Jenifer, and husband Atora, a District Agriculture Officer for Goroka Department for Agriculture and Livestock (DAL) have been active in promoting and distributing african yams to help local farmers (also producer–sellers, but farmers plant on a bigger scale) to prepare for climate change. In 2003, Atora and Jenifer bought eight seeds and in 2009 started distributing them on behalf of the DAL. They planted and multiplied them: 'After six years of planting, we were just producing them and ate some then planted more, but were just producing, nothing more', explained Atora. However, Atora had a feeling that they must keep doing this and look after these yams carefully – he believes this was a premonition. In 2009, the couple made their first sale:

> Two white men were looking for yam seeds everywhere and when they heard a couple were growing them in the Eastern Highlands, they contacted us and took forty tubers (five or six seeds in each tuber). They gave us 400 kina (150USD).

This was their first encounter of people willing to spend a large amount on their yams. A second order of yams brought them 1,400 kina (528 USD) and the third, 40,000 kina (15,100 USD). By 2011, Atora and Jenifer made two sales of 27,000 kina (10,192 USD) and 45,000 kina (16,987 USD) to people around the country wanting to buy african yams – a considerable amount to be made from a single food product for one family.

'Even people who never gardened are now planting african yams', said Atora. Atora and Jenifer explained that the production of white yams has been beneficial for food security as people are able to guarantee themselves food during times of drought. The tubers continue to grow even when there is a shortage of water and they do not require any additional inputs, other than labour. They can also be stored inside for long periods of time and become a lucrative commodity when other crops fail.

The virtues of african yam have contributed to it gaining socio-cultural value in the Eastern Highlands, thousands of miles away from Nigeria. I realised this one dry, dusty weekend as I sat underneath a temporary tarpaulin shelter in a suburb of Goroka town. As I chatted in hushed tones with the relatives of a man who had recently passed away, and been a District Officer for DAL, his colleagues arrived with a truckload of african yams. Atora and Jenifer had collected these from a group of local producers they had been training as part of their projects for adaptation to climate change. It was set up that the DAL would buy the yams from the producers and then redistribute them as seed for interested buyers and other growers.

In this case, the DAL used the african yams as a gift for their colleague's family, giving them as a gesture of their 'sori' or shared grief. The tubers were piled up outside the tarpaulin shelter in front of a crowd of crying women sat cross-legged on a woodchip floor. As well as the yams, the

DAL staff presented the family with an envelope of money that was a contribution towards food and drink for the mourners who would sleep there for the coming days and nights.

I asked Atora what the colleagues would have usually given at such a time, had they not had a large amount of african yams. He said it would be their custom to bring particular kinds of bananas, taro and sugarcane. These are all highly valued traditional foods that are essential at major events such as bride wealth and death payments. As a local village leader once told me, even if a huge amount of money was being given from one clan to another, still sugarcane, taro and bananas must be included. It is remarkable then that african yams were considered an acceptable gift in place of taro and bananas, having enough of a collectively recognised value to not only show shared sorrow but also offer a kind of compensation to the family of their deceased colleague and friend.

I have heard other such accounts of african yam in ceremonial exchanges and seen them included in the mounds of food given on special occasions. It is an expensive food item to purchase in the market and rarely found there as surplus seems to be recirculated as seed. The fact that it takes skill and knowledge to grow and is fairly hard to come by, has contributed to the african yam becoming a valuable incorporated into socio-political, ceremonial exchanges. This may change as more and more people produce the crop, for example, sweet potato is not considered a valuable food in the ceremonial economy as it grows profusely and frequently, thus being eaten daily. Without the state's agricultural agency promoting adaptation to droughts, the african yam may never have made an entrance into the ceremonial economy. So climate change as a discourse that has forewarned people of climatic instability has shifted the parameters of the food crops that can be exchanged in ceremonial and socio-political contexts.

It is not only new crops that grow as a result of being promoted by such actors, the language and awareness of climate change is also spread in the process. For example, Jenifer made a television programme demonstrating the ways in which WIA were promoting local food production to prepare people for climate change. Climate change then becomes an issue that people react to, changing what they produce and exchange as a result of information provided by government sponsored agencies. What people grow is not just affected by climate change itself, it is also affected by the way people talk and react to the discourse of climate change. This is worth thinking about when considering how people receive and interpret global discourses. It also shows the way in which local forms of exchange can adapt and incorporate new food crops, despite not being 'traditional' or historical like crops such as winged bean or taro. How the socio-cultural value of these crops is assigned is another matter.

LOCAL SPACES, POWER AND 'FRICTION': CLIMATE CHANGE ENTERING LOCAL DISCOURSES OF SOCIAL CHANGE

Discourse (as language and ideas) makes its way from global levels into local contexts. Goroka is affected by global shifts and social trends, but people in these local contexts also shape discourse to suit their understanding and experience (Lahsen, 2004; Rudiak-Gold, 2011; Tsing, 2005). They also experience the reality of dramatic weather changes and what that can mean for their day-to-day survival.

'Climate change' is a commonly used term in PNG thanks to local radio, information provided by agricultural agencies and warnings issued by the government. This is one way in which people are framing their experience of change generally, and in relation to food. They may not know the science behind climate change, but they can relate to it in their experience of real change in their ecological and socio-economic environments.

However, not everyone who uses the term climate change is actually referring to the weather patterns and seasons, as an externally driven global phenomenon. One local elder, George, uses the term to refer to environmental transformation that is occurring closer to home. He told me he thinks it is the 'cash money system that is destroying the environment'. George believes gradual changes in behaviour in Goroka have led to a damaged local environment, telling a particular story of the deforestation of Mt. Agopeya. This local mountain is under stewardship of his clan but people living on the peripheries of Goroka town have been chopping down trees growing on it for firewood. Firewood, as with food, has increased in price over the last three decades and as more people live in peri-urban areas and struggle for an income, they venture into the local hills to find low-cost energy sources. George believes there is a growing lack of respect for the environment, one aspect of rapid urbanisation.

George explained that he first heard the term climate change after the international Rio conference in 1992 (the Earth Summit); he was invited by logging activists to attend a follow up Pacific conference in Balolo, PNG. The fact that George, a local landowner, teacher and artist, is aware of global conferences about climate change and has engaged in forms of environmental activism against deforestation is just one example of the way people in this 'frontier culture' engage with global discourses and ideas (Sassen, 2007; Tsing, 2005).

He relates his own observations of the damage that has resulted from a growing urban population to the global processes that have been used to explain anthropogenic climate change, that is, unsustainable consumption of the earth's resources. As Henrietta Moore argues that, however, knowledge is now at once local and global but this does not necessarily make it universal (Moore, 1996; Sassen, 2007). George is keen to make Mt Agopeya a protected area, open to tourists as a means of conserving his ancestral environment and generating an income for the community. He told me he plans to plant trees that have been lost during 'development' in PNG and make the area open and appealing to visitors. This is what he sees as a solution to the pressures that urbanisation and population growth have put on the local environment, and thus a solution to local 'climate change'.

CONCLUSION

The food landscape in Goroka is changing and has been since, and before, first contact with Europeans in 1933. Today it is being affected by the ecological consequences of climate change, the discourse of climate change and the impacts of broader social change within an emerging market economy. When people in Goroka use the term climate change to explain why there are no longer exchanges across ecological zones, they do not always acknowledge the change in preference and action of growers alongside the ecological changes. Although real change in seasons and temperatures has occurred, other social factors have contributed to the decline in exchange of traditional foods such as 'winged bean'. Nevertheless, such social exchange has certainly not come to an end. As shown in the case of african yams, people continue to perform their obligations and reciprocal relations with one another, sometimes using new crops that have gained status and prestige. In this case, climate change has affected what foods are being grown and exchanged in a different way, not through ecological change, but through being a global discourse that has led to agricultural agencies encouraging adaptation. This has also meant a lucrative income for those initiating the circulation of african yams.

Highlanders are once again seen to be adapting to change. People in Goroka are accommodating to the recent changes in temperature, finding new opportunities in both market and social exchanges (yields of coffee have increased throughout the year and coastal crops are now being grown in EHP). This demonstrates that rural people in the Pacific are not inevitably victims of climate change but find ways to adapt when they can (Barnett and Campbell, 2010). However, at the time of writing, friends in Goroka reported experiencing heavy rainfall, hail storms and floods that led to landslides destroying people's houses and gardens. People are struggling to grow major cash crops such as English Potatoes due to the increasing number of pests. And there is also a rise in malaria cases (and insufficient health care to cope with this), broadly accepted by health professionals and locals alike to be the result of rising temperatures. Some changes are difficult to adapt to.

'Climate change' in Goroka is also a synonym for change generally. When climate change is referred to, it often means something that is not within local people's control. For many, this is the case as severe weather events, which were never prompted by them (or their consumption), have a major impact on their livelihoods. Yet the decline in social exchange of local food crops (such as winged bean, karuka and marita), across time and space, due to a change in local preferences and practices also makes people less able to cope when problems of extreme weather or poor harvests arise. Ceremonial food exchange does continue to be important to people in Goroka and a continuation of this practice may contribute to the resilience of rural and urban people in the face of environmental and socio-economic challenges.

NOTES

1. I have translated all original Tok Pisin (the lingua france of PNG) accounts into English.
2. This type of data exists for Goroka from 2000 to 2014, however, it is held at the Goroka weather station and was in the process of being analysed and released at the time of writing.
3. As found out through personal contact with staff members.
4. The urban setting of the Goroka market made this research inevitably multi-ethnic; the province has 26 different languages spoken within its boundaries. In order to speak with people throughout the province about their food production and climate change, I had a comprehensive understanding of Tok Pisin (TP) as this is how people from different language groups speak with each other. I spent the

majority of my time with people who are related to the Gahuku tribe and reside in the Zauka valley. Although many of these people speak Gahuku as their first language they all, except a few very elderly women, could speak the lingua franca Tok Pisin.

5. Free education up to secondary school policy was introduced in 2014, where only project fees should be paid for by the family of the pupil.

6. Experts who give advice to farmers through visits and trainings in villages.

7. She pronounced this in English rather than a Tok Pisin version that is used in coastal parts of PNG: 'kilamet senis'.

8. Two types of Pandanus, the former bearing nuts and the later producing red seeds that are boiled and squeezed into a bright red oily liquid, *Pandanus conoideus* and *Pandanus brosimos*.

9. However I did hear winged bean still commonly used in communities closer to Kainantu where the tradition of communal production for exchanges continues to be strong.

ACKNOWLEDGEMENTS

Significant thanks go to the numerous families and communities in Goroka that welcomed me and told me their stories, showed me their gardens and shared their food. I could not have conducted this research without the permission and support of Papua New Guinea's National Research Institute in Port Moresby and the support of the Marsden Grant, Royal Society of New Zealand. Since returning from Goroka, I owe many thanks to Mark Busse, Cris Shore, Robin Hide, Richard Eves, Geir Presterudstuen, Yasmine Musharbash, Jane Horan, Samuel Taylor-Alexander, Fraser MacDonald and Correlli Barnett, as well as the anonymous reviewers, who gave me substantial feedback and suggested edits during the writing of this paper. Thanks go to Briar Sefton, University of Auckland for illustrations.

REFERENCES

Akin, D., & Robbins, J. (1999). An introduction to melanesian currencies: Agency, identity, and social reproduction. In D. Akin & J. Robbins (Eds.), *Money and modernity: State and local currencies in Melanesia*. Pittsburgh, PA: University of Pittsburgh Press and Association for Social Anthropology in Oceania.

Asian Development Bank. (2013). *Climate change to hit Papua New Guinea economy hardest in the Pacific*. Retrieved from http://www.adb.org/vi/news/climate-change-hit-papua-new-guinea-economy-hardest-pacific-adb. Accessed on May 21, 2015.

Barnett, J., & Campbell, J. (2010). *Climate change and small island states: Power, knowledge and the South Pacific*. London: Earthscan.

Barnett-Naghshineh, O. (2014). Asbin: A has been of socio-political exchange? An analysis of changing values in the highlands, PNG.

Bashkow, I. (2006). *The meaning of whitemen: Race and modernity in the Okokaiva cultural world.* Chicago, IL: University of Chicago Press.

Benediktsson, K. (2002). *Harvesting development: The construction of fresh food markets in Papua New Guinea.* Denmark: Nordic Institute of Asian Studies.

Bourke, R. M., & Harwood, T. (2009). *Food and agriculture in Papua New Guinea.* Canberra: ANU Press.

Brookfield, H. C., & Brown, P. (1963). *Struggle for land.* London: Oxford University Press.

Brouwer, E. C., Harris, B. M., & Tanaka, S. (Eds.). (1998). *Gender analysis in Papua New Guinea.* Washington, DC: World Bank Publications.

Busse, M. (2012). Negotiating market and community: The (Im)morality of reseller transactions in the goroka market, Papua New Guinea. Domestic Moral Economy Workshop, Canberra, September.

Curry, G. (1999). Markets, social embeddedness and precapitalist societies: The case of village tradestores in Papua New Guinea. *Geoforum, 30,* 285–298.

Denham, T. P., Haberle, S. G., Lentfer, C., Fullagar, R., Field, J., Therin, M., … Winsborough, B. (2003). Origins of agriculture at Kuk Swamp in the highlands of New Guinea. *Science, 301,* 189–193.

Ellis, D. (2003). Global averages, local extremes. In S. Strauss & B. Orlove (Eds.), *Weather, climate, culture.* Oxford: Berg.

Epstein, T. S. (1982). *Urban food marketing and third world development: The structure of producer-seller markets.* London: Croom Helm.

Errington, F., & Gewertz, D. (1999). *Emerging class in Papua New Guinea: The telling of difference.* Cambridge, MA: Cambridge University Press.

Escobar, P. (1995). *Encountering development: The making and the unmaking of the third world.* Princeton, NJ: Princeton University Press.

Eves, R. (2000). Waiting for the day – Globalisation and the apocalypticism in Central New Ireland, Papua New Guina. *Oceania, 71*(2), 73–91.

Fajans, J. (1988). The transformative value of food: A review essay. *Food and Foodways, 3,* 143–166.

Finney, B. R. (1969). New Guinean entrepreneurs; indigenous cash cropping, capital fromation and investment in the New Guinea Highlands. *New Guinea research bulletin.*

Fiske, S. J., Crate, S. A., Crumley, C. L., Galvin, K., Lazrus, H., Lucero, L., … Wilk, R. (2014). Changing the atmosphere. Anthropology and climate change. *Final report of the AAA global climate change task force* (p. 137). Arlington, VA: American Anthropological Association.

Fleming, E., & Hardaker, B. (2010). *Pacific 2010: Strategies for Melanesian Agriculture for 2010; tough choices.* Australia: National Centre for Development Studies.

Hide, R. (1984). *Environment, society and history.* Boroko: Institute of Applied Social and Economic Research.

Hughes, P., & Sullivan, M. (1990). Climatic change and agricultural production in the highlands of Papua New Guinea. In J. C. Pernetta & P. J. Hughes (Eds.), *Implications of expected climate changes in the South Pacific region: An overview* (p. 270). UNEP Regional Seas Reports and Studies.

Igoe, J., Neves, K., & Brockington, D. (2010). A spectacular eco-tour around the historic bloc: Theorising the convergence of biodiversity conservation and capitalist expansion. *Antipode, 42*(3), 486–512.

Ives, S. (2014). Uprooting "indigenity" in South Africa's Western Cape: The plant that moves. *American Anthropologist, 116*(2), 310–323.

Jacka, J. (2009). Global averages, local extremes: The subtleties and complexities of climate change in Papua New Guinea. In S. A. Crate & M. Nuttall (Eds.), *Anthropology and climate change: From encounters to actions* (pp. 197–208). Walnut Creek, CA: Left Coast Press.

Kahn, M., & Sexton, L. (1988). The fresh and the canned: Food choices in the pacific. *Food and Food Ways: Explorations in the History and Culture of Human Nourishment, 3*(1–2), 1–18.

Keil, D. E. (1974). *The inter-group economy of the Nekematigi, Eastern Highlands District, New Guinea.* Ph.D. thesis, Northwestern University. Retrieved from http://library.ucsd.edu/dc/object/bb60763711

Kyakas, A., & Wiessner, P. (1992). *From inside the women's house: Enga women's lives and traditions.* Brisbane: Robert Brown.

Lahsen, M. (2004). Transnational locals: Brazilian experiences in the climate regime. In S. Jasanoff & M. L. Martello (Eds.), *Earthly politics, worldly knowledge: Local and global in environmental politics.* Cambridge, MA: MIT Press.

Lipset, D. (2013). The new state of nature: Rising sea-levels, climate justice and community-based adaptation in Papua New Guinea (2003–2011). *Conservation and Society, 11*(2), 14.

Maclean, N. L. (1988). The commoditization of food: An analysis of a maring market. *Canberra Anthropology, 12*(1–2), 74–98. 161–179.

Meggit, M. (1974). Pigs are our hearts! The te exchange cycle among the mae enga of New Guinea. *Oceania, 44*, 165–203.

Meig, A. (1984). *Food, sex and pollution: A New Guinean religion.* Ann Arbor, MI: Rutgers University Press.

Moore, H. L. (1996). The changing nature of anthropological knowledge. In *The future of anthropological knowledge* (pp. 1–15). New York, NY: Routledge.

Monbiot, G. (2009). Climate displacement has begun – But hardly anyone has noticed. *The Guardian.* Retrieved from http://www.theguardian.com/environment/georgemonbiot/2009/may/07/monbiot-climate-change-evacuation

Munster, P. (1986). *A history of contact and change in the Goroka Valley, central highlands of New Guinea, 1934–1949.* Ph.d. thesis, Deakin University.

National Statistical Office of Papua New Guinea. (2011). *'Preliminary figures' national census of housing and population – Papua New Guinea.* Retrieved from https://spmt.files.wordpress.com/2010/09/preliminary-figures-census-2011-1.pdf

Orlove, B., Lazrus, H., Hovelsrud, G. K., & Giannini, A. (2014). Recognitions and responsibilities: On the origins and consequences of the uneven attention to climate chamge around the world. *Current Anthropology, 55*(3), 249–275.

Rappaport, N. (1968). *Pigs for the ancestors: Ritual in the ecology of a New Guinea People.* New Haven, CT: Yale University Press.

Read, K. (1965). *The high valley.* New York, NY: Charles Scribner's Sons.

Risimeri, J. B., Gendua, P. A., & Maima, J. B. (2001). The status of introduced white yam in PNG. In R. M. Bourke, M. G. Allen & J. G. Salisbury (Eds.), Food Security for Papua New Guinea. *Proceedings of the Papua New Guinea food and nutrition 2000 conference,* PNG University of Technology, Lae 26–30 June 2000. Australian Centre for International Agricultural Research, Canberra, pp. 783–787 (online at ACIAR).

Rudiak-Gold, P. (2011). Climate change and anthropology: The importance of reception studies. *Anthropology Today*, *27*(2), 9–12.

Sassen, S. (2007). *A sociology of globalisation*. London: W.W. Norton & Company.

Sexton, L. (1986). *Mothers of money, daughters of coffee: The Wok Meri movement*. Ann Arbor, MI: UMI Research Press.

Sillitoe, P. (2001). Pig men and women, big men and women: Gender and production in the New Guinea highlands. *Ethnology: An International Journal of Cultural and Social Anthropology*, *40*(3), 171–192.

Sillitoe, P. (2010). *From land to mouth: The agricultural economy of the Wola of the New Guinea highlands*. New Haven, CT: Yale University Press.

Strathern, A., & Stewart, P. (1999). Objects, relationships and meanings. In D. Akins & J. Robbins (Eds.), *Money and modernity: State and local currencies in Melanesia*. Pittsburgh, PA: University of Pittsburgh Press.

Strathern, M. (1988). *The gender of the gift: Problems with women and problems with society in Melanesia*. Oakland, CA: University of California Press.

Strauss, S., & Orlove, B. S. (Eds.), (2003). *Weather, climate, culture*. Oxford: Berg.

Sullivan, M. (1990). The effect of sea level rise on raised coral and high islands. In J. C. Pernetta & P. J. Hughes (Eds.), *Implications of expected climate changes in the South Pacific region: An overview* (Vol. 128, pp. 218–223), Nairobi. UNEP Reg.Seas Reports/ Studies.

Sullivan, M. (1991). The impacts and projected climate change on coastal land use in Papua New Guinea. In D. Lawrence & T. Cansfield-Smith (Eds.), *Sustainable development of traditional inhabitants of the Torres Strait region* (pp. 33–58). Cairns: Great Barrier Reef Marine Park Authority.

Sutton, A., Mountain, M. J., Aplin, K., Bulmer, S., & Denham, T. (2009). Archaeozoological records for the highlands of New Guinea: A review of current evidence. *Australian Archaeology*, *69*, 41–58.

Tsing, A. (2005). *Friction: An ethnography of global connection*. Princeton, NJ: Princeton University Press.

Vásquez-León, M., West, C. T., & Finan, T. J. (2003). A comparative assessment of climate vulnerability: Agriculture and ranching on both sides of the US–Mexico border. *Global Environmental Change*, *13*(3), 159–173.

Waiko, J. (1993). *A short history of Papua New Guinea*. Oxford: Oxford University Press.

West, P. (2010). Making the market: Specialty coffee, generational pitches, and Papua New Guinea. *Antipode*, *42*(3), 690–718.

West, P. (2012). *From modern production to imagined primitive: The social world of coffee from Papua New Guinea*. Durham, NC: Duke University Press.

FARMERS' RAINFALL ANTICIPATION: INCIDENCE AND PATTERNS IN WESTERN NIGERIA. ADVANTAGES OF FOCUS AND PROBLEMS OF EXTRAPOLATION IN CASE STUDIES

Jane I. Guyer

ABSTRACT

Purpose – *This paper examines localized conditions and responses to what people see as ordinary variations in the* weather, *drawing on their own archive of knowledge and practice for "coping" with it, as distinct from year-to-year* climate *patterns that may entail "adaptation."*

Design – *This paper draws on ethnographic field research and rainfall statistics collected in 1968–1969 and 1987–1988, in a rural area of Western Nigeria where guinea-savannah small-scale farmers now grow increasingly for the market. Research in the 1980s was designed to track all changes since the 1960s. It is revisited here to draw out the rainfall variable.*

Climate Change, Culture, and Economics: Anthropological Investigations
Research in Economic Anthropology, Volume 35, 135–153

ISSN: 0190-1281/doi:10.1108/S0190-128120150000035006

Findings – *In the 1980s, farmers noted a decline in the first rains of the early growing season, and a change in the short dry season, over a period of three years, in a way that differed from the expected patterns of twenty years previously. The shift is confirmed by rainfall statistics. Their crop repertoire choices are noted.*

Limitations and research implications – *The paper's themes are culled from a broader range of observations over the 20 years. The interweaving of the variables in complex change over several decades is noted as a research challenge.*

Originality – *Local time series, interpreted through the local archive of social and technical practice, offers a rich entry point into what the recent AAA climate change review refers to as coping and adaptation, with respect to what I call "weather" and "climate."*

Keywords: Inter-annual rainfall patterns; Nigeria; farmers; archives for coping

INTRODUCTION: WEATHER AND CLIMATE; COPING AND ADAPTATION; AND ANTICIPATION AND PREDICTION

One implication of the concept of "global climate change" is that, as the recent AAA report on *Anthropology and Climate Change* (Fiske et al., 2014: section 5) notes, there are yet finer-grained levels from which to learn than the global and the regional at which increasingly clear and stabilizing climate change may have particularly powerful effects. These would be in "community-centered," "place-based," and "path-dependent" studies. Each locality may be particularly difficult to generalize from, with respect to such a micro-factor as people's perceptions of the changeability of the weather and the crop-task-market expertise that works with it. However, the processes themselves of working back and forth from current situations to past experience may be provocative from one case to another. Local peoples' ordinary anticipations of changeability, their sedimented knowledge archive, and capacity to anticipate, then brought to bear on a perceived shift in the parameters of knowledge itself, is the topic that drives this paper.

At its most comprehensive, the focus on climate change is understandably on global phenomena such as general temperature change, rising oceanic water levels, and a changed frequency of violent events such as typhoons and floods, all of which are thought to have anthropogenic causes at the global aggregate level. Until the recent consolidation of a detailed global scientific record, sudden phenomena such as floods and famines that deeply disrupted social and economic life may often have been assimilated, by experts and the people alike, to cosmological understandings of the interconnection of the moral, spiritual, and material worlds. Indeed, the etymology of "climate" refers to a macro-property: the angle of the earth to the sun (*klima* in classical Greek), so setting the predictable parameters of the annual cycle in the length of daylight, warmth, and rainfall that are ultimately governed – from the human vantage point – by the movement of the sun in the sky. There are, inevitably, also variations *within* these predictable parameters that can be anticipated as to their nature, but not predicted as to their precise magnitude and duration in any particular year or place. In English, we refer to the latter as "weather," whose etymology derives from "storm" and "wind."[1] Weather is always changeable over the short run, and from year to year. In what self-consciously conceptualized sense this is understood by different populations and their experts, in relation to the broad parameters of climate, is a function of the framing of ecological knowledge, local experience, and expertise. How knowledge of local weather, and the archive of experience for dealing with it, may come into play when people's perceptions shift toward longer-term climate change, is a question that anthropology may be particularly well-placed to address.

In mundane times and domains of life, a capacity to anticipate and track the possible and likely changeability factors in rainfall and temperature (as "weather"), within broad schemes of predictability (as "climate"), would be necessary for all the pragmatic purposes of productive life: the development of the crop varietal repertoire (with characteristic maturational conditions and durations), the setting of the annual task cycle, the invention of the toolkit, and the development of soil technologies. Thus, the means for anticipating the possibility of localized and temporary variations, within the climate parameters, become a part of the range of knowledge. The question I address here, provoked by fieldwork observations in 1988, in giunea-savannah West Africa after a phase of catastrophic drought in the Sahel, is whether, and when, peoples whose livelihood depends on understanding and adapting with alacrity to the weather, begin to question the possible stabilization of this changeability into a repetitive pattern from year to year, the kind of new climate parameters that climate change theory expects to develop. If so, they may see this shift as more challenging than dealing with the weather, and as meeting up with other macro-changes in their lives such as the growth of markets. And yet they may also see their existing weather-responsive repertoire as a strong resource to deal with the shifts, to be developed much further. If so, local observations and recourse to their own archives of knowledge and practice might produce ideas for others.

I will use the concept of "anticipation" to refer to farmers' orientation to expected changeability: expected in its incidence but known to be highly variable in its particular manifestations, over any time period from one day, or one week, or one year to the next. Farmers know, for example, that first planting in almost any nonirrigated system could be followed by rainfall shortage that wastes the grain, or by sudden deluges that wash it away. There have to be back-up modes of dealing with both eventualities. They may not be mobilized very often, nevertheless people have cultivated modes of improvisation, and have developed a willingness to hedge against many small risks, for example, by saving seeds even at planting time and by adjusting the crop repertoire when a major crop would be lost if it could not be dried in the sun immediately after harvest. There are also micropolitical means whereby dangers can be shifted temporarily from one social category to another, which would then reflect and reinforce inequalities and vulnerabilities if their range moved into a novel and more regular mode.

The AAA report (Fiske et al., 2014) is worth quoting at length on this point, where it differentiates among these tactics, distinguishing coping from adaptation:

> Adaptations are part of cultural knowledge and practice, evolved over time; in effect, part of the overall toolkit for life. Coping refers to decision-making in novel situations for which there is no ready culturally integrated institutionalized response. Both adaptation and coping involve problem solving and decision making but coping involves improvisation for immediate versus long term goals. Outcomes of both adaptation and coping may be replete with winners and losers, and may have unanticipated results. (2014, pp. 42–43)

When it comes to production and making a living, they see "adaptations" as based on a large archive of technical and social knowledge that has been curated, passed on and made available for mobilization in possibly new combinations for situational "coping." I suggest here the importance of learning how such archives of knowledge are composed and mobilized. This is where ethnography is so important, since the sources may be very subtle and interconnected, and only small traces may persist in the documented records, and even in the daily life of the period of field research. And they may work differently for different sub-groups with different accessibilities to the knowledge and resources to be configured. Anticipations of the weather are based on experience and extrapolation. For climate questions, such microstudies of the weather may be difficult to carry out with respect to the past, since it is usually the major events that leave a strong trace in both the records and people's memory. Dealing with the weather can seem too mundane, although historical ecological studies may not yet have exhausted all the possible source materials that could be interrogated with an ethnographic imagination, including our own field research records. The recent work of Mavhunga (2014) on Zimbabwe makes the more general case for the scientific basis of African memory-archives of technical and spiritual expertise with respect to the environment in which people have lived for long periods.

THE CASE IN POINT

In this paper, I reexamine a group of Western Nigerian small farmers' observations and reactions in the 1980s, when three successive years of a late start to the first rainy season seemed to them to have set in after the major droughts and famines in West Africa in the early 1980s. These farmers live in a guinea-savannah area, outside the forest belt but well south of the Sahel. Their livelihoods depend on the production of food crops, mainly for supply to the large and growing urban populations of Southern Nigeria: Lagos, Abeokuta, Ibadan, and other smaller towns. Rainfall is bimodal, so there are two growing seasons, with an expected short dry season in July and August, which is crucial for the drying, transformation, and storage of crops from the first season, with some rain allowing for planting for the second rainy season. Then there is the long dry season, from December to early March, when a great deal of the land preparation is done, for the whole coming year.

At that time in the 1980s, I was undertaking research on changes in production since the late 1960s, mainly attendant on the expansion of the urban markets. This unusual incident of rainfall variation is mentioned in my book (Guyer, 1997) but not as a moment to pause over, to work outwards into the evidence on the adjustments and changes that might well have been, in part, skillful accommodation to a variability that was anticipated as a happening but not in its specific nature. The farmers' immediate perceptions of a possibly emergent pattern, at this specific moment can offer a focus for opening up the question of the archive of knowledge, and the processes of consultation and choice, that I raised in several other publications (Guyer, 1992, 1996; Guyer & Lambin, 1993; Lambin & Guyer, 1993), but have not yet brought together in relation to understanding microecological dynamics that may well be exacerbated by cumulative climate change.

I revisit my own evidence, with the thought that other anthropologists' records may well contain more sources for thinking about responses to perceived novel patterns in the weather than we have yet brought to the fore, and may nuance our approach to adaption and coping as concepts for processes increasingly thought of as parts of climate change.

WEST AFRICAN AGRO-ECOLOGICAL STUDIES AND THE KNOWLEDGE "ARCHIVE"

For West Africa, an ethnographic effort gained momentum in the 1970s aimed at appreciating the dense richness of agro-ecological knowledge in intercropping systems, on forest-savannah borders, in response to crops and varieties imported from the New World in the early modern period, in response to policies that attempted to graft monocropping into these systems in the colonial period, and where hunting and gathering remained part of the range of the local culinary and livelihood strategies. My own work in rural Western Nigeria focused primarily on responses to the new kinds and levels of market demand as urban populations grew rapidly and gave the rural areas increasing characteristics of urban hinterlands (Guyer, 1997).

In that period of the 1960s through the 1980s, scholars of West African agricultural systems learned enough to know that we were only at the beginning of a vast excavation enterprise to find the depth and variety of ecological knowledge from which the management of the anticipated changeability of the weather draws. In recent research into money management issues (especially debt and the timing of repayment), a Yoruba elder gave me an image that could appositely apply to knowledge more broadly. He drew my attention to asha generally translated as "custom," as the source they draw on in order "not to make things worse." Karin Barber (personal communication) offered me the etymology for asha from the verb "to choose," so referring not to a coherent body of established principles and practices so much as to the process of invoking and selecting from past, accumulated compilations of experience. By implication, asha is retained in all its multiplicity, and then consulted, situationally, for sources appropriate to the problem of the moment. Working back into agro-ecological knowledge, then, this image may guide an approach back into our own field research records, which may contain more of this archive of knowledge, and its situational mobilization, than we were fully aware of at the time, when climate change – as distinct from incidental events and just weather – was not a large presence in the intellectual climate of the times.

Anthropological and ecological studies in the last decades of the twentieth century focused precisely on intricate varietal specializations and intercropping patterns, which had been closely accommodated, over centuries: to very small micro-ecologies within regional patterns and to the expectation of unpredictability. Harlan argued that "Crop innovation for adaptation to rainfall is identifiable very early in African agricultural history, in varietal improvement in cereal production that maximizes the use of limited moisture" (Harlan, 1980, p. 339). Richards (1983) has compiled the sources on the particularly rich and complex resource-use traditions on the savannah-forest border, which has been relatively densely populated for centuries, and to the peoples' integrated knowledge of the whole life-world, with its complex interdependencies of species: flora and fauna, wild and domesticated, in all their seasonally shifting relationships. In his comprehensive review of African ecological studies, he pointed to rainfall unpredictability as a particular challenge: "Rainfall reliability and erosion risks (especially as a result of intense convection storms at the beginning and end of the rainy season) are significant issues throughout the tropical zone." (1983, p. 24). He drew close attention to the great range of crop varieties, especially in the rice production zone, and suggested that the introduction of monocrops and high-yield varieties under colonial and postcolonial agricultural policy could be a problem for farmers working in ecologies where "(r)isk minimization is generally a more important criterion for small farmers than profit maximization. Intercropping, then, is one of the great glories of African science" (1983, p. 26), infused by "the scarcest of all ecological resources – human ingenuity" (1983, p. 56).

Rainfall variation is a recurrent theme, and Richards even suggests that the amounts, timing and synchronization of all aspects of production strategies are "mathematized." African farmers have "extremely highly developed mathematical ability to match tasks and available labor, to time operations in relation to climatic events," (1983, p. 51) through "human ingenuity" (1983, p. 56). This was shifting, as he wrote, in the current political and economic context. The complementarity of farming with hunting and gathering may have been in decline, due to broader shifts in land control. The postcolonial period saw the rise of monocrops, and the decline in the varietal repertoire for old crops, which appear to have been quite dramatic in the postcolonial period, especially with the policy promotion of high yield varieties of commercial crops (see Bassett, 2001, pp. 129–130, on cotton in Cote d'Ivoire). The main challenge for farmers in this case was the mobilization of peak period labor in new ways, to manage the larger harvest. While Bassett also points out the effect of rainfall variation on the timing of seasonal labor mobilization, he does not trace out in detail whether, and with what implications, certain changes that are quite similar to the one on which I focus here, played out through the process by which people mobilized the archive of knowledge and practice, and combined it with possible novel components, especially if they see certain aspects of the anticipated variability of the weather taking on new characteristics.

In his critical appreciation of Richards' (1983) review, Michael Watts drew attention to the particular centrality of rainfall unpredictability in the more arid regions, and especially in stratified polities and under conditions where the poorer farmers could not affect policies. With "farmer responses to rainfall variability in northern Nigeria, every farming season was a massively risky undertaking for some households and many of the strategies to mitigate environmental unpredictability were simply not options in the farmer repertoire" (Watts, 1983, p. 78). This was not yet a concern in my own field site, with its still-open land frontier and before the expansion of mid-scale farming, but it would certainly become a concern under new circumstances.

Focusing on the management of the intersecting temporal regimes of cropping cycles, labor mobilization, and rainfall, which have to be brought into ongoing productive convergence, offers an important empirical entry point into this ingenuity, as both archive and process, and with new crops. It also creates space for thinking back to framing and scaling issues in the understanding of the dynamics of change and adjustment across different temporal ranges in Western Nigeria and Africa more broadly, of the kind that I worked on together with specialist in remote sensing, Eric Lambin (Guyer & Lambin, 1993; Lambin, 1997), and with a group of other scholars in both disciplines (Guyer, Lambin, Cliggett, & Walker, 2007). If, as I heard from farmers in the 1980s, particular shifts seemed to have become more frequent and more problematic, we can try to follow out the perceptions and implications.

RAINFALL SHIFTS IN IBARAPA: 1960s AND 1980s COMPARED

The farming cycle in the 1960s was crafted to a dependable bimodal rainfall pattern: two growing seasons (March to July; late August to November) and two dry seasons (December to March; mid-July through August), with crop maturational cycles and intercropping options that had been adjusted to the temporality of the rainfall. Fig. 1 and Table 1 summarize the rainfall for 1966–1968, the conventionalized cropping seasons, and the areas devoted to each crop. The largest areas were devoted to crops most dependent on the regularity of the rains, especially in the germination phase immediately after planting: egusi-melon, maize, and yams. Cassava was primarily a last rotation in the four-year cultivation cycle, persisting into the beginning of what was conventionally eight years of fallow, and thriving regardless of the rainfall distribution. The other crops were all on smaller plots, with planting adjusted to the mid-season pattern of rainfall.

Egusi-melon was one of the most lucrative crops. A valued dietary element in all of southern Nigeria, and a commercialized crop for at least a hundred years, it had been sold into the large settlement areas. And it grew well in the relatively sandy soils. Indeed, the name of the area itself, Ibarapa, is said to derive from the word *bara*, which designates the vine on which the melon grows. The growth and maturation cycle corresponds to the length of the first rainy season, and the labor-intensive harvest, seed extraction and drying can be undertaken during the short dry season in July. The field was then planted in guinea corn for developing the stalk-trelliswork for yam planting in November, or in maize. Other fields were intercropped with crops whose harvests could be done carefully enough not to cause damage to the others: cowpeas, tomatoes, peppers, and eventually cassava, planted every 10 rows, and then taking the whole field at the end of the four-year cycle. Cassava was the other leading commercial crop, requiring facilities for processing and also a dry season for making into various flours.

One can see how precisely these crop succession and intercropping patterns corresponded to a bimodal rainfall pattern. In the 1960s, no-one could tell me of a time when the harvest had failed. This population had never known penury or impoverishment in the variety of their diet. During the years for which I could gather precise rainfall data, from the Nigerian Tobacco Company Offices from which they worked with contract farmers, the intra-annual rainfall variation patterned very closely with these seasonal cycles, even though the total precipitation varied widely from year to

Table 1. Growing Periods for Certain Crops, with Planting and Harvest Times.

Crop	Jan.	Feb.	Mar.	Apr.	May	Jun.	Jul.	Aug.	Sept.	Oct.	Nov.	Dec.
Egusi		P	—	—	—	—	H					
Groundnut			P–	—	—	–H						
Maize			P	—	—	—H		P	—	—	—H	
Cowpeas				P–	—	—	—	H				
Tobacco				P	—	—	—	H P	—	—	—H	
Tomato						P—	—	—	—H			
Peppers	—	—H				P	—	—	P—	—H	—	—
Guinea Corn							P—	—	—	—	—H	
Yam	—	—	P—	—	—	—	—	—	—H		P—	—
Cassava	—H	—	—	—	–P–	–H–	—P	H—	—P	—	–H–	—

Note: P = approximate planting date; H = approximate harvest date.

year. The rains fell consistently at the beginning of the growing period, which was essential for establishing the crop, and they fell sharply during the short dry season for the harvest and drying. Only 1968 was sharply different, but then only in the second rainy season, when the rainfall was simply much heavier. This second season period was beyond my time in the field, but the crops planned for that season would not have done badly under heavy rainfall conditions during the growth period. Fig. 1 shows the monthly rainfall for those years.

The totals were 840.86 mm (1966), 573.34 mm (1967), and 1,368.79 mm (1968), but one can see clearly how the first growing season followed closely the expectation of gradually rising, or at least stable, rains after planting and during the main growth period, and in 1966 and 1967 the short dry season was perfectly predictable. In 1968, the first harvest would have been difficult due to heavy rains, but to know the depth of the effect on labor mobilization and drying one would need to know the distributions at a yet more fine-grained level, in the *daily* patterns of rainfall, in addition to the monthly totals. The effects of one very heavy storm might be more easily mitigated than trying to navigate harvesting and drying over a continual lighter rain that was consistent over two weeks. Under the former conditions, egusi-melon seeds could be sheltered during a storm, and then put back out to dry. Under the latter conditions, they would start to germinate which would deeply affect their marketability. It would also affect income sources for the women, one of whose main revenues – paid in kind so

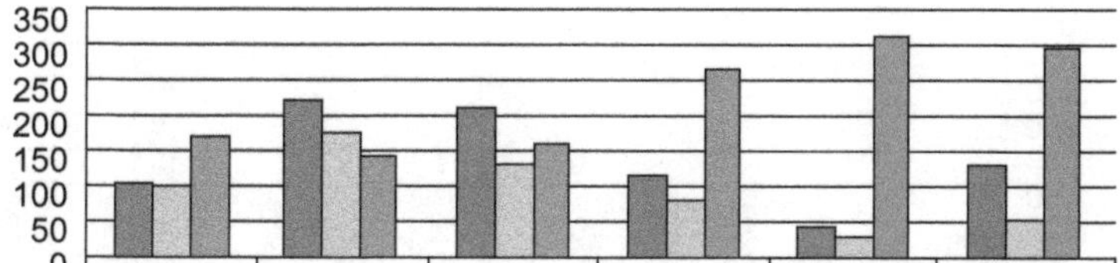

	April	May	June	July	August	September
1966	106.35	225.63	212.69	115.48	44.67	136.04
1967	97.97	179.95	136.29	80.71	30.96	47.46
1968	175.89	144.92	161.68	269.04	317.01	300.25

Fig. 1. Rainfall in mm, April to September, 1966–1968. *Source*: Nigerian Tobacco Company at Igbo-Ora, daily measures. All farmers with whom I worked farmed within 10 miles of this station.

manageable through strategies of one's own choice – was their pay for an intense period of harvest duties.

So, in the 1960s, the *total* rainfall varied widely from one year to the next, but it followed the same rising and falling pattern in all three years. The season I followed in detail showed that the areas under specific crops, the intercropped cultigens, and the income followed quite closely the expected routines of the rainfall cycle, and the patterns of cultivation show the confidence that farmers had in a traditional crop whose success depended on their capacity to anticipate. Egusi accounted for 41% of the acreage in the first season (see reported in Guyer, 1997, pp. 174–175), all the small-scale farmers whose work I studied grew at least some, and it accounted for 25% of their cash income for that season. In the 1980s, when I returned for a restudy in the context of commercial growth, the intra-annual rainfall distribution seemed to the farmers to have shifted, while the total rainfall was quite stable from year to year.

It was an apparent shift at the beginning of the first season that farmers were noting in 1988, and seeing it as a new phenomenon: new in the sense that it had recurred. In all three years, the rainfall in April was lower than 20 years previously.

The table in Fig. 2 shows the rainfall for 1986–1988.

In 1986 and 1987, the April rains fell far short of the 100 mm, which it far surpassed in 1966 and 1968, and fell only very slightly short on in 1967. Even in May, at the height of the growing season, rainfall barely

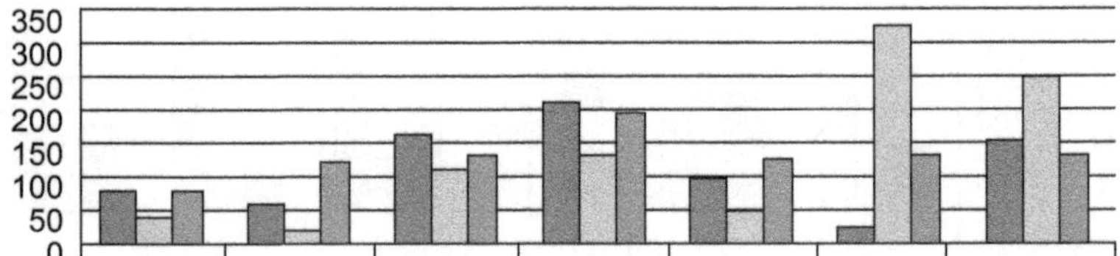

	March	April	May	June	July	August	September
1986	74.9	58	166.6	212	98.5	20	155.6
1987	40.6	21	111.1	135.9	48.4	323.7	252.3
1988	81.5	117.8	131.7	192.9	123.5	133.3	137.5

Fig. 2. Rainfall in mm, March to September, 1986–1988. *Source*: Nigerian Tobacco Company at Igbo-Ora.

approached the 150 mm that had been always – or almost (1968) – surpassed in the previous period. Then it dropped off in July, just as in the past, in all three years. In effect, the whole first season had been retarded by a month at the beginning, but not correspondingly lengthened at the end. It ended up being very difficult to manage for planting, since they had always planted with the first rains (usually in March) and been able to count on the seeds germinating. If they had no rain, the planted seeds dried and were perhaps eaten by predators, and the field had to be planted all over again, which required either a solid back-up plan for seed storage, or enough funds to buy seed in the market. As long as such rainfall was coincidental, the volume stored and the management of access might fall into what we have called an archive of adaptations: intercropping, interseasonal storage by at least some members of any localized community, and various situational adjustments in crops and labor. But if it became predictable, especially under intensified commercialization, when saving seed would necessarily reduce money income, then it was likely that the classic cropping patterns would have to be modified *from the outset* of the farming season, which would then change the whole farming year, including forms and rhythms of labor mobilization.

It was the repetition of this rainfall pattern that was disturbing to the farmers in 1988. This was no longer seen as a single year variability in the weather, even though it could still be dealt with through existing means, for example, by the redistribution of the seed bank among close coresidents in the farming villages, and kin. But this would be difficult to do year after year. It had now emerged as an apparent repetition; lower rainfall in April was happening every year. And although there had been one very wet August and September in 1968, by 1987 and 1988, the higher rainfall in the short dry season was being repeated. At the time, the farmers wondered whether it might be a result of the droughts of 1970s and the early 1980s, and if so, that the climate was changing, beyond what they anticipated with weather fluctuations of the kind we can see in the data for the 1960s. If they planted any of the traditional crops with the first rains, as was the custom in the past, they risked losing the first planting, and if they did not, they risked the harvest, because the rainy season no longer seemed to coincide with the crop-maturation process. In either case, much more of the crop would have to be stored against possible adversity, and therefore less would be sold. By my measurements, the total acreage under egusi had fallen to 23%, with only 40% of the small-scale farmers growing it, in 1987, and by 1988 providing only 3% of their income (Guyer, 1987, pp. 174–175). The largest contribution to acreage and income became cassava, which is relatively weather resistant and can be harvested at any time. In 1968–1969, cassava was grown by 81% of farmers, on about a quarter of their acreage, yielding 18% of their cash income. In 1987, this was 100% of the farmers, over 40% of the cropped acreage, yielding 35% of their cash income. The only cash source that even approached this level was from intensive cultivation of tomatoes, at 27% in 1988.

Rainfall was not the only factor in these shifts, although the farmers did mention the new rainfall pattern as a problem. There had been other shifts in the cropping and labor cycles since the 1960s, largely due to transport and market conditions, which would shift the crop–rainfall synchronies into two other rhythms: toward the center of the year, May to September (tomatoes and peppers), and a growth in crops with year-long (and more) cropping cycles (cassava). Neither of these would be as affected by the rainfall changes as would the traditional crops. All these changes are documented in chapter 11 (Guyer, 1997), entitled "The Small Scale Male Farmers in 1988." As noted above for the aggregate, in each of the four sites where I had undertaken field research in both periods, the proportion of farmland devoted to the traditional anchor crops, egusi and yam, had diminished significantly, and the proportion devoted to tomatoes and peppers, and cassava, had risen. The proportion of farmers devoting any land at all to each of the crops had changed in the same direction: far fewer growing egusi, yam, and even maize, while far more were growing tomatoes, peppers, and cassava. Cassava and tomatoes now accounted for a much higher proportion of income than in the past. Farm size had expanded, due to the use of tractors for clearing land and the increasing supply of migrant labor, which now came not only from central Nigeria but also from neighboring Benin and Togo.

Women had also entered farming, on their own account, after the transport revolution of the 1970s oil-boom, so they could be somewhat less dependent on payment for their work in the egusi-harvest on the farms of the men. Their own preference for cassava, which they had always processed for market, added to the overall movement of the crop profile to this most climatically resistant of crops. However, they also did grow egusi, whose harvesting labor was familiar from the past: washing, drying, and eventually shelling. So in 1988, they devoted a higher proportion of their smaller farms (less than half the size of men's farms) to egusi (30% in the first season) than did the men (23%).

I returned again in the 1990s and early 2000s, working on a different field research topic, after almost another 20 years, to find that the increased availability of transport, and rapidly growing market demand from Lagos, had encouraged yet newer crops, such as fresh watermelon. The regional market hinterland could shift its practices in yet new ways as a result of improved access, although the changed rainfall patterns in March and April were still a challenge to be navigated, and in new ways now that the traditional rotations had already been somewhat shifted. The great advantage of watermelon for the farmers – like tomatoes, peppers, and leaf vegetables in 1980s – has been its consistent demand in the urban markets, and the fact that no processing at all is required. The rapid transport to market is taken care of by the large supply of vehicles and traders, and the accessibility of the cities by improved roads. Of course, selling fresh crops then deprives those whose livelihoods had rested on processing of one of their opportunities for earning a living. The one problem with watermelon, however, for the growers, is that it is an early season crop and can be very vulnerable to exactly the disturbances in the early rains that farmers had been puzzled by in 1980s. Since the Nigerian Tobacco Company no longer operated, accurate rainfall data are not available to me, but people's reports now focused on damaging deluges and storms rather than the lateness and low level of the early rains. One year when I visited, the surge into watermelon had greatly diminished due to the new risks of turbulence, rather than simply diminution, in the early rains. The mid-year pattern was altogether less problematic, which seemed to have encouraged maize and other short-season crops.

In brief, the micro-dynamics of rainfall have proven crucial to crop and labor change. Farmers have been accustomed to anticipating the possibilities of variation in the weather for a very long time, but dislocations emerge when they perceive the shifts becoming patterns rather than events, and where adaptive practices of the past have to face the coping challenge of novel crops, and of having to be actualized at the beginning of the season. From the weather of the 1960s, which was changeable but within easily anticipated parameters, farmers had moved to an apparently novel pattern in the 1980s, when they began to shift the crop repertoire in favor of crops not requiring April rains and August drought. And then, with new commercial crops, they began to focus on the apparent increase in sudden events such as damaging storms. Anticipation necessarily presumes possible projections, so we can ask how experienced repetitions, which we might call patterns, affect the way in which the archive is used, and – over time – how it is reconstituted, even in its most nuanced combinations of knowledge and practice.

INFERENCES: MICROECOLOGIES, SMALL FARMERS, AND PERCEIVED CLIMATE CHANGES

How the various forces of change – the climate, the market, crop varietal shifts, changes in labor patterns – might be seen in convergence with one another is a very complex question with deep implications for the qualities on which recent anthropological work on climate change has focused: vulnerability, adaptation, resilience (AAA, 2015). We can see, however, how these particular small-scale farmers have continually adjusted by both drawing on the archive of knowledge and practice for situational response to events and adding to it with new crops for anticipatory projection of patterns. As Richards (1983) suggested, there has been a clear capacity to keep up the reconfiguration of crops, cycles, and conditions. But it may well be shifting again, where the crops that can survive the perception of new rainfall patterns, and also maintain a market niche, are taking more of the total acreage. As a result, diversity, intercropping, and the maintenance of a complex archive for situational adaptation may be declining. More and more farmers are falling into the category I called, in 1988, "sideliners": that is, people planning and expecting to make their income-earning careers in other occupations, and for whom a farm is a strategy for additional cash income earning, rather than for self-provisioning, as in the past when all artisans had diversely cropped farms to feed themselves and their families. Logic suggests that the farms of sideliners would take on a speculative and perhaps monocropped mode, responding to rapidly shifting market possibilities, and worked by rented tractors and hired labor. Managing the vicissitudes of the climate, and/or simply the weather, may then move into the sphere of monetary credit and debt, with less of the diversity and adjustable complementarities of 1980s. Having hired a tractor and laborers for land preparation, they have to be paid. The rainfall pattern can then be threatening to *monetary* relationships. Coping is now a strategy with monetary relationships rather than primarily with an ecological-adaptive archive. Some men seem to have lost confidence in farming as an occupation, unless the conditions of production and the market can be more firmly assured, and their other income-earning occupations can work well enough to support a dependable money income. The last innovations I heard about were thoughts on expanding orchard crops that produced preservable fruits, beyond the patches of cocoa that were already yielding well in certain villages in the 1960s and locust-bean (a gathered commodity). Special attention was being given to cashews (for the nuts). All these adjustments are built on certain skills at synchrony, although now more clearly a matter of synchronizing money and markets than a wide diversity of provisioning crops adjusted to the rainfall.

In several papers, and my book, I have argued for closer attention to the changing shape of diversity, under the many shifting conditions of the late twentieth century in Africa, as referenced already in the text. Here, I have delved a little more deeply into my sources on rainfall variation, worked on farmers' original perceptions of change in 1988, and looked at the logics that one may infer from the patterns of their activities and the conditions of their livelihoods. Clearly, several historical dynamics are intersecting at once, of which perceived shifts in weather patterns can be one. It can be a particular challenge to discern the interactive dynamics of the different forces at work, but farmers' perceptions of the rainfall–crop interaction can be a particular entry point into their understanding of change: as simply changeable weather conditions manageable by the archive of common sense, or as climate change, in both the literal and the metaphorical senses: for actual rain, for the climate of the economy and the climate of social life.

As a final step, I would indicate the problem of extrapolation from micro-regional data to regional and broader processes. In a conference paper, we (Lambin & Guyer, 1993) addressed the analytical problem raised by the AAA report on climate change with respect to scales of analysis, and extrapolation from localized studies to higher levels, whether those dynamics might be intermittently recurrent phenomena or examples of possibly novel processes that are becoming more widespread and more frequent due to larger climatic change. We argued for the necessity of being attentive to three kinds of region: ecological-demographic regions, economic regions (such as urban food supply hinterlands) and socio-cultural-political regions (we suggested the French term *terroirs*, or community regions) where there are commonalities of resource control and local knowledge. In doing this, we can be experimental in our thinking about possible patterns, as perceived from data or from farmers' responses, and also modest enough to know that our evidence is fragmentary, and all the more so in its temporal features: all the different temporalities that have to be brought together to make a living, and the historical shifts in the context, of which rainfall change is one that would affect the practices in all three analytical frames.

By seeking out a specific moment, such as the puzzling rainfall shifts of the 1980s in Ibarapa, one can move closer to the immediacy of action. The experience of episodes and the logics of calendrics can become entry points into emergent systemic dynamics, some of which may well be shaped by several forces of which climate change is only one. In one sense, adaptation is the mobilization of common sense. What can be seen with one lens as a relentless, even tedious, rhythm of life, from another is a series of accidents waiting to happen and downward spirals waiting to gain momentum, all of which can be mitigated. Here, I am reminded of Paul Richards' argument that in many agricultural systems just staying in place is a major achievement of human ingenuity. By shifting the frame of reference, in accordance with several plausible analytics, and in accordance with people's own landmarks for improvised "coping" that may turn into "adaptation," we can prepare the terrain for helping certain seeds of observation to germinate into much larger investigations that may be relevant to understanding and mitigating the implications of climate change more broadly than in the field where the observation first occurred. This would be especially the case when we can analyze the intersection of weather, climate, crop repertoire and political economic change, to see how they come together and are separated by both the people and the analyst. In my own case here, what started as an agro-ecological concern by the farmers began to be mitigated by shifts in the crop repertoire and modes of farming for an expanding market, but then became a monetary concern as the seasonality of money and the seasonality of work have to come together in new ways.

NOTE

1. Etymonline.com Consulted May 23, 2015.

ACKNOWLEDGMENTS

A first version of this paper was written for a panel organized by Naveeda Khan for the third annual conference of the Initiative on Climate Adaptation Research and Understanding through the Social Sciences (ICARUS), 2012. My field research was supported by the National Institute of Mental Health (1968–1969) and the National Science Foundation (1988). I acknowledge the importance of suggestions by Donald Wood and two reviewers.

REFERENCES

Bassett, T. J. (2001). The peasant cotton revolution in West Africa. *Cote d'Ivoire, 1880–1995*. New York, NY: Cambridge University Press.

Fiske, S. J., Crate, S. A., Crumley, C. L., Galvin, K., Lazrus, H., Lucero, L., ... Wilk, R. (2014). Changing the atmosphere. In *Anthropology and climate change. Final report of the AAA Global Climate Change Task Force*. Arlington, VA: American Anthropological Association.

Guyer, J. I. (1992). Small change: The rhythms of labour and 'Leisure' in a Western Nigeria Savanna town, 1969–1988. *Africa*, *62*(4), 465–489.

Guyer, J. I. (1996). Diversity at different levels: Farm and community in Western Nigeria. *Africa*, *66*(1), 71–89.

Guyer, J. I. (1997). *An African niche economy. Farming to feed Ibadan 1968–88*. Edinburgh: Edinburgh University Press, for the International African Institute.

Guyer, J. I., & Lambin, E. F. (1993). Land use in an urban hinterland. Ethnography and remote sensing in the study of African intensification. *American Anthropologist*, *95*(4), 839–859.

Guyer, J. I., Lambin, E. F., Cliggett, L., & Walker, P. (2007). Temporal heterogeneity in the study of African land use. Interdisciplinary collaboration between anthropology, human geography and remote sensing. *Human Ecology*, *35*(1), 3–17.

Harlan, J. (1980). The tropical African cereals. In D. R. Harris (Ed.), *Human economy in Savanna environments* (pp. 335–343). New York, NY: Academic Press.

Lambin, E. (1997). Appendix A. Remote sensing study of the Agrarian landscape in idere. In J. I. Guyer (Ed.), *An African niche economy* (pp. 235–237). Edinburgh: Edinburgh University Press.

Lambin, E. F., & Guyer, J. I. (1993). The complementarity of remote sensing and anthropology in the Study of Complex Human Ecologies. African Studies Association Meetings. Boston University Working study series, Working Paper Series No. 175.

Mavhunga, C. C. (2014). *Transient workspaces. Technologies of everyday innovation in Zimbabwe*. Cambridge, MA: MIT Press.

Nigerian Tobacco Company. Rainfall Records, Igbo-Ora. Oyo State.

Richards, P. (1983). Ecological change and the politics of African land use. *African Studies Review*, *26*(2), 1–72.

Watts, M. (1983). "Good try Mr. Paul". Populism and the politics of African land use. *African Studies Review*, *26*(2), 73–83.

CLIMATE CHANGE POLICIES AND ADAPTIVE BEHAVIOR OF FARMERS IN SOUTHERN CHINA

Meine Pieter van Dijk and Hao Li

ABSTRACT

Purpose – *The paper analyzes the adaptive behavior of farmers in the Yunnan province of China, where drought is occurring more frequently. We focus on the experiences with adaptation to climate change by farmers in the rural areas of China.*

Methodology/approach – *The research is based on a survey and a number of in-depth interviews of key stakeholders in a drought-stricken region.*

Findings – *Where the government is not always coming forward, the farmers take initiatives to adapt to the new situation of drought. Different mechanisms are being used, some linked to government policies and subsidies, other initiatives are initiated by the farmers themselves, individually or in small groups.*

Research implications – *More research on the livelihood strategies is necessary to better understand what these strategies mean for the household income and hence for the survival chances of poor households.*

Climate Change, Culture, and Economics: Anthropological Investigations
Research in Economic Anthropology, Volume 35, 155–175

ISSN: 0190-1281/doi:10.1108/S0190-128120150000035007

Practical implications – *Climate change encourages local actors to play a role in drought adaptation, developing policies for mitigating the consequences of drought, trying to create water markets and involving local companies and water user associations. The research suggests stimulating the initiatives of the farmers and to create an enabling environment for them.*

Social implications – *Without government policies we will see growing inequalities in the rural areas of China.*

Originality/value – *We studied how in the case of drought farmers react to adapt to the new reality. Different adaptation strategies are distinguished and their relation to different government policies is established. We observed that farmers find their own solutions and create their own governance structures to assure for example supply of additional water to their fields.*

Keywords: Drought; adaptation mechanisms; farmers; governance structures; climate change policies; climate change adaptation

INTRODUCTION

The Chinese government has initiated policies at national level to deal with climate change risks, especially drought. However, are farmers benefiting from current government policies concerning climate change and which initiatives have they taken themselves to deal with droughts. In this contribution, we will assess on the one hand to what extent initiatives started by the government to deal with climate change are actually working through at the local level. On the other hand we will analyze the initiatives of farmers in Yunnan province to deal with the more frequent occurrence of droughts due to climate change. After analyzing the top-down approach by the government, the initiatives of the farmers will be analyzed and the role of government policies in these initiatives will be assessed.

After an overview of the relevant literature, the research strategy and the institutional context will be presented. An analysis of the initiatives taken by the farmers shows their skills to adapt to a new situation of more frequent and more severe droughts, more or less independently from the government policies in place.

LITERATURE REVIEW

Climate change influences the biophysical, sociocultural, and socioeconomic context of rural households, and creates unprecedented external stresses for livelihoods (Nelson et al., 2009; Paavola, 2008; Selvaraju, Subbiah, Baas, & Juergens, 2006). The major effects on livelihoods are reduction of existing livelihood options, and perhaps more importantly in the short to medium run, a greater volatility and unpredictability in streams of livelihoods benefits (Agrawal & Perrin, 2008).

Traditionally, the mainstream of research on building up rural adaptive capacity to drought focuses on identifying possible strategies which could be applied by farmers. These strategies include crop diversification (Benioff, Sandra, & Jeffrey, 1996; Chiotti, Johnston, Smit, & Ebel, 1997; Downing, Ringius, Hulme, & Waughray, 1997; Parry, Rosenzweig, & Iglesias, 1998), changing cropping intensity, crop mix, type and location (Smit & Skinner, 2002), constructing water infrastructure (Smith & Lenhart, 1996), crop insurance (Goodwin & Smith, 1995), diversification of income earning, and employment opportunities (Barrett, Reardon, & Webb, 2001). However, the empirical research reveals that the abilities to access to these strategies are different across rural households (IPCC, 2007; Keshavarz et al., 2010). Against this background, Li et al. (2013), Li, Gupta, and Van Dijk (2013) have identified and evaluated the effectiveness of the strategies taken by rural households to cope with drought with the help from governments, based on a household survey at local level along the Lancang River. Different adaptation strategies were identified and will be classified at the end of this paper.

RESEARCH QUESTIONS AND STRATEGY

This research contrasts national top-down policies with grassroots level initiatives of Chinese farmers in the Yunnan province. The region was selected because it is known to be a dry region, and more frequent droughts are the expected result of a gradual process of climate change, and many policies have been implemented at Yunnan province. An analysis of the multilevel governance structures in place shows on the one hand that the implementation of policies and programs frequently gets stuck at the provincial or local level. However, on the other hand farmers and local enterprises are no passive by-standers, and take initiatives to deal with the many challenges they are facing, which are partially climate change related and in particular concern the increasing incidence of droughts. This research focused on the Lancang river area, within the Yunnan province, where seasonal droughts have become more important.

Within the Yunnan province the Upper Mekong River Basin was taken as the case study area. Two districts and six villages were selected there.[1] The study area is mountainous and it makes a big difference whether the villages are located close to the river or up in the mountain, where it is more difficult to bring the water from the river. We found some of the villages have successfully made the transition from cultivating grain to growing cash crops like tea, coffee, or bananas, while in other villages the inhabitants have to supplement their agricultural income by revenues from wages in the city, where one or sometimes more members of the household work for a short period in the dry season in the construction sector, or for a longer period in the manufacturing industries. To do justice to the different conditions we selected, in two districts (Pu'er and Xishuang), villages near the river, uphill, and in between and some of the richer villages which had made the transition toward exportable cash crops and some of the poorer villages, which are depending heavily on economic activities outside the village.

The research is based on a survey and a number of in-depth interviews of key stakeholders in this drought-stricken region. In total 80 households in six villages were interviewed. The villages are scattered in mountain area along the Upper Mekong River. The distance between two villages is usually very far. Traveling from one village to another is time consuming. Also doing the interview is very time consuming. Sometimes assistance from an interpreter is necessary as some of the minorities living there have their own languages. In addition, in order to try to find the linkages between policy and farmers' adaptive strategy, the interviewers tried to find the story behind the drought adaptation strategy selected, which also requires additional time. In order to improve the representativeness of the sample, we used the website "Digital of rural Yunnan."[2] This website offers the statistical data on biophysical and socioeconomic status of all 124,206 natural villages in Yunnan from 2006 onward. The natural village or community level is interesting in China because the people living there have the freedom to elect their own village leaders, who are not paid by the government but tend to be able to negotiate with the authorities at the administrative village level for favors, for example related to mitigating the negative consequences of drought.

The interviewed villages were selected according to altitude and main source of income of the village. We wanted villages along the river and higher up in the mountains. Similarly, we included villages which have modernized agriculture (coffee and tea growing) and villages depending on traditional agriculture and other sources of income (because of rubber plantations, tourism, or in the cities). In the selected villages, we asked the head of village to recommend 10 households ranging from successful to poor according to their income.

The questionnaire used in the survey provides a good entry point to discuss the issues with rural households. In a number of cases the heads of households or key informants were willing to provide additional information and additional questions were asked. Additional information focused on the strategy of the household, trying to find out to what extent its members adapt to the more frequent and severe droughts in this area.

THE PHYSICAL AND INSTITUTIONAL CONTEXT

The Yunnan province is one of the underdeveloped regions in the southeast China. In Yunnan province, 5.55 million people lived below the national poverty line in 2008 ($176 per capita) (Li, 2010), accounting for 13.8% of the numbers of the poor in China (Zhang, 2011).

To achieve sustainable livelihoods is always a challenge in Yunnan province, even if there is no climate change. The region is mountainous, hence transportation is difficult, and infrastructural facilities are limited. Although this region is rich in natural resources and has a high natural productivity, it suffered from serious poverty caused by relative isolation from outside markets and underdeveloped local markets.

In the Yunnan province, climate change has increased potential risks, including agricultural drought, strong wind and hail, soil deterioration and erosion, grass land degrading, significant reduction in runoff, and a loss of biodiversity (YPG, 2008), which turned the local social and environmental system into a more fragile one. Among the potential risks, drought and concomitant water shortage not only significantly constrain farmers' ability to get out of poverty trap, but even bring some people back into poverty (Ye et al., 2012; Zhou, Lü, Xu, Zhang, & Jiang, 2011). Water in rural areas plays a key role in irrigation, in raising livestock, and in manufacturing activities such as brewing and brick-making, and hence is the nexus of rural household production and income earning (Calow, MacDonald, Nicol, Robins, & Kebede, 2002).

Most national policies have been translated into local policies, but local governments have also developed policies based on their own goals. We looked at policies trying to improve the adaptive capacity of the farmers. For example, guidelines were developed for the local government what to do in the future. However, no detailed action plans have been prepared to implement these climate change policies. This process could not start before laws and departmental rules were issued and were followed by a series of ordinances (Fig. 1). The problems are that the departmental rules are binding for different departments and there were policies that overlap and even conflict with each other. Sometimes there were conflicting interests between institutions (Li et al., 2013).

The Chinese governance structure for dealing with the effects of climate change can be described as a multilevel governance structure for drought management. There are at least seven levels of administrations in China, which are nation, province, municipality, county, township, administrative village, and natural village. The first five levels have constructed strictly hierarchical bureaucracy system to implement administration orders from upper government. The cadres of local government will be appointed by upper government, while they are authorized to choose their own developing pathway under the guidance of central government.

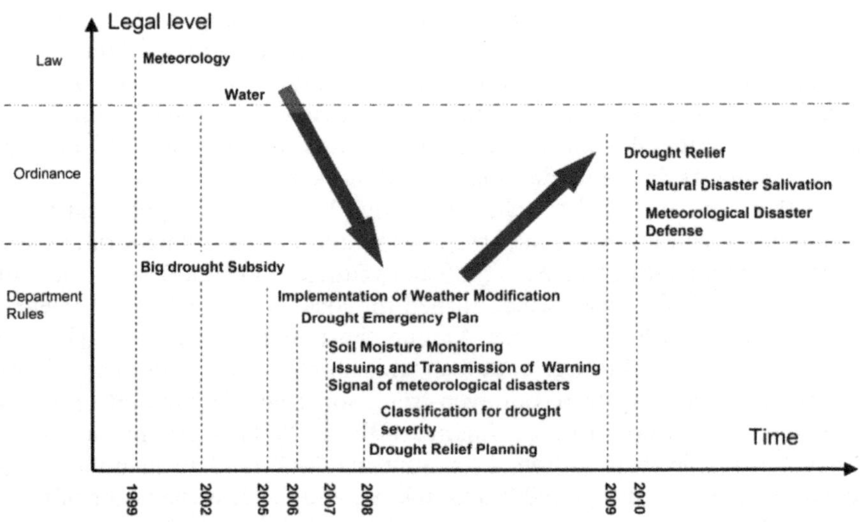

Fig. 1. Drought Policy Shifts.

The county is the core and a powerful unit of regional development, because issuing land rights is appropriated to this level of government. This institutional arrangement also affects other types of resources allocation. The administrative village level is the lowest Chinese level of officially paid officials. The cadres of administrative village are elected by villagers and appointed by government. The natural villages (or the dispersed communities) are loosely under the control of administrative village and the elected leaders are not officials. The natural villages are more autonomous and are usually held together by informal institutions such as race and lineage.

The policies from central government are hardly implemented at the lowest level of the natural village or the community level. The reform of agricultural taxing system has made this even more difficult. Before 2001, public funds mainly come from taxes (*tiliu*), fees (*fei*), and apportionment (*tanpai*) in poor areas (Tsai, 2002). After 2001, the central government started to reduce farmers' burden, and this reform is finished with a milestone of abrogating agricultural tax completely since 2006. Although this reform has improved farmers income, it also has blocked an important source of public funds, which was used to offer public goods and services to farmers as well as to promote bureaucratic agencies mediating between government and farmers at the natural village level. With the government retreating from the village affairs, the traditional approach of government will be substituted by polycentric governance structures. This leads to more self-organization of farmers and companies, which now play a more prominent role in the autonomous villages (Rogers & Hall, 2003).

In 2006, a comprehensive assessment report on climate change has argued that placing more emphasis on regional case studies on adaptive measures can reduce the uncertainty caused by limited knowledge of climate change and has identified several pertinent adaptation measures for main sectors. These measures have been formally brought forward in 2007 National Climate Change Program. A report on national policies and activities coping with climate change was published each year, which aims to sum up the performance (Talbot, 2005). In 2009, the Standing Committee of the National People's Congress has passed a resolution for coping with climate change, which points to the need to improve laws and regulatory systems as well as improving public participation in the future.

Drought relief policies, which aim to resolve the disaster following the drought, include the instruments of early warning, weather modification, drought relief emergency response plan, soil moisture monitoring, and drought relief planning. These policies are top down and are usually implemented by government directly with the purpose of helping victims out of disaster. The National Meteorology Law in 1999 and the revised National Water Law 2002 have provided principles and frameworks of measures in coping with drought. Under these two frameworks, different functional departments in central government have crafted their own policies separately based on self-responsibilities. After several years of implementation, it began to be realized that overlaps and gaps simultaneously exist in these policies. As a result of reconsideration, the more powerful national ordinance was used to integrate separated policies. These national ordinances include the National Drought Relief Ordinance in 2009, the National Natural Disaster Salivation Ordinance, and the National Meteorological Disaster Defense Ordinance in 2010 (see Fig. 1).

Capacity-building policies, which aim to enhance the adaptive capacity of farmers, include improving water availability and use efficiency, disseminating drought relief knowledge and technology, enhancing farmers' income, initiating drought hazard loans and insurance, and strengthening social network of local communities.

The amendment of the National Water Law (2002) is a milestone of changing from "supply management" to "demand management" in the dominant integrated water resources management paradigm. This amendment puts more emphasis on the utilization of property rights, market mechanisms, and water saving technology to improve water use efficiency to resolve impending water scarcity. Water levies for direct withdrawals has ended free water use since 2006. One of main ideas of policy is to offer direct help to farmers which could improve their capacity to deal with the consequences of drought. The fund for special projects to promote advanced technology as well as subsidies for improved seeds and drought relief falls under this national policy. Other policies include:

(a) Dissemination of drought relief knowledge and technology by promoting the diffusion of agricultural technology, drought-resistant seeds, new seed planting techniques, water conservation technologies, and technology promotion projects;
(b) Improving farmers' income through temporary jobs in urban areas, allowing land use rights transferring,[3] conversion of farmland to forests, subsidies for improved seeds, and obtaining the necessary equipment;
(c) Drought hazard loans and insurance such as subsidized loans from the Council for farmers to reduce poverty, post-drought emergency

low-interest loans to farmers; and drought disaster insurance financed from a central budget.

Since the policy process is so complicated and uncoordinated, the government constituted comprehensive coordinating committees in the State Council including representatives from relevant ministries, administrations, banks, scientific bodies, social associations, NGOs, and even the army. Four such committees relevant to drought management are the State Council Leading Committee on Poverty Alleviation and Development (STLCPAD), the National Headquarters of Flooding Control and Drought Relief (NHFCDR), the National Leading Committee of Disaster Alleviation (NLCDA) and the State Council Leading Committee of Coping with Climate Change (SCLCCCC).

In addition to the four local leading committees corresponding to the national leading committees, there are four other leading committees at the local level in charge of reforming the small rural water infrastructure, providing subsidies for improved seeds, ensuring labor transfers from rural to urban areas, and converting farmland into forests in mountain areas. These committees coordinate the drought management of different departments, and use joint meetings for making decisions. There is a mismatch if certain committees are not present at the local level, or when the national level is not involved, as for example in the case of the leading committee on reforming of small rural water infrastructure and the leading committee on Grain for Green program.

Most policies are paternalistic, designed to help farmers but do not involve them in decision making. Relations between the province and local level are not as hierarchical as between the national and provincial level. The decision-making authority is at the local level, especially the county government. Central government's funds and subsidies are the two main incentives for local government to implement related policies.

At the provincial level, there is a general program and assessment report on climate change. In 2010, Yunnan was selected as one of the five Green GDP priority provinces for experimenting with the climate change program. This will bring Yunnan province more political and financial support from central government and it will provide the cadre of the Yunnan province more opportunities. The Yunnan government has issued drought relief ordinances that predates those of central government and issued a meteorology ordinance that corresponds to national legislation. Because of these two ordinances, all levels of local government have been asked to craft annual plans for weather modification, drought relief planning, and emergency plans. A policy initiated by Pu'er government has involved mobile phone companies which offer short message service (SMS) for farmers to spread information about droughts. These services are very important, but not always freely available.

Many national level initiated policies are not made specific after the provincial level, namely planning for water saving, improving agricultural and water resources technologies promotion, raising money and labor on a case-by-case basis, management reform of subsidized loans to reduce poverty, regulations on subsidies for agricultural machinery, and regulation of water-withdrawal permits and levies. Policies not yet translated to the provincial and local level are central government policies of funding of water and sanitation services, the policy on low premium insurance which asks for a 50% contribution from provinces, the policy of promoting rural microfinance, the policy on subsidies to small rural water infrastructure construction, and policies on creating WUA.

To improve the adaptive capacity, five groups of policies have been emphasized by all levels of governments, which include reform of the agricultural technology promotion system, management reform of Small Rural Water Infrastructure, transferring free labor from rural to urban areas, subsidies for improved seeds, and conversion of farmland to forest in mountain areas. There are two reasons why these policies have been implemented at all levels. First, because local governments have received subsidies or project fund from central government. Therefore, some special activities are necessary. The other reason is that these policies give more freedom to local governments to implement them.

There are still many challenges faced by climate change governance in China. The national policies can be characterized as emphasizing drought relief policies more than capacity building. The underlying approach seems to be "disaster management" (Wilhite, 2005). The government prefers response systems to deal with disasters. Secondly, the government mostly promotes subsidies and fund-based approaches and has only just started introducing some market measures. The funds and subsidies can help government to achieve its objectives rapidly. However, these instruments are not flexible and constrained by government's budget. Meanwhile, the effects can be limited by corruption.

Initiatives of the Farmers

Due to certain drivers we observed a shift in the governance structures in China (Li & van Dijk, 2012), resulting in more opportunities for participation and local initiatives. Two major drivers of change in China are the rapid economic growth taking place, in particular, in the eastern part of the country and the process of urbanization. Recently, climate change and a growing awareness of environmental degradation have helped to create an environment in which governance is shifting. Governance is also shifting because of a different environment, economic incentives, and urgent events such as climate change and a growing awareness that another approach is needed than the top-down and command and control approach. The shifts in governance are facilitated by more emphasis on decentralization (van Dijk, 2012).

We will now analyze the initiatives of the farmers and what all these policies mean at the community or natural village level. The social capital of local communities is enhanced through the establishment of local drought relief teams, Water Users Associations (WUA), rules for communities to raise money and to provide labor for infrastructure works. From 2002, the central government has realized that investments and maintenance of the water infrastructure depended mainly on the government. Because of the limited budget of the government it cannot meet the increasing needs for more water infrastructure. Therefore, the central government is promoting reforms that encourage local governments, companies, or private individual to do investments in middle or small-scale water infrastructure works, through approving transfers of property rights of these water infrastructures from government to nongovernmental actors. Although the big water infrastructures are traditionally regarded as controlled by central government, the forms of management are more diversified and decision making is more polycentric. This reform shows the effects of decentralization of water governance. It leads to more participation in water affairs, encourage multiactors to take responsibility, and improve the public awareness on water conservation. According to the household survey, several strategies have been chosen by farmers to deal with droughts. Their essence will be analyzed and the link to government policies will be identified.

S1 Engage in Nonagricultural Work

A total of 19% of the adults (18- to 60-years old) were engaged locally in nonagricultural work during the year preceding the survey. About 14% of all adults migrated to other places for nonagricultural temporary jobs last year. Information on jobs and improved seeds is mainly disseminated through neighbors, lineages, and relatives. It was also found that several farmers came back to their village during the last five years, for marriage, old age, or because of the low income in urban area or to invest their savings and engage in more capital intensive agriculture.

Although local governments have implemented special policies to offer training courses or job information to farmers, nobody among interviewees has ever been involved in this kind of projects.

S2 Livestock or Fish Culture

Increasingly households become involved in livestock or fish culture diversifying their sources of income, although fish breeding is not very popular in this area, because it requires land, which is located near a rich source of water. However, 46 out of 60 (75%) households raise livestock, and 20 out of these 46 (43%) households earn money from selling their livestock. There was always a tradition of growing animals for the spring festival.

Increasingly the farmers also do these activities for the market because since a number of years the government subsidizes female pigs and increases the fertility of the pigs by providing growth hormones.

S3 Rotation, but Still Growing Grain

To reduce their dependency on irrigation water, farmers cultivate crops requiring less water, for example maize instead of rice. On about 18 of the 148 plots used by the households interviewed (or in 12% of the cases), the farmers have resorted to changing the rotation pattern, partly because there was not enough irrigation water. Changing the crop structure is a risk reduction strategy, which is related to the "Grain for green program," which aims to conserve soil and water in mountainous areas.

The subsidy for planting grains from central government aims to secure China's food supply. In Yunnan province, this subsidy tries to stimulate farmers to plant crops which are less vulnerable to drought. The planting association and technology-promoting agencies play a role in implementing the policy.

S4 Change Crop Structure

Farmers who change from planting maize to tea in mountainous areas will receive 40 USD per mu per year. About 45 out of 60 (75%) households have planted trees, and almost all of them consider it a measure to improve their income. However, in that case they do not get the grain production subsidies. Coffee is more vulnerable to frost and gales than to drought, farmers prefer to plant coffee instead of rice or maize because the frost and gales seldom occur here. In order to improve farmers' ability to plant economic crops, special planting associations supported by local government have been established to disseminate related technology. These associations are usually organized by technology-promoting agencies on behalf of local government and they often organize some training course for farmers. The organization usually takes the administrative village as the unit and farmers can choose to join or not. About 56.7% of the households have planted coffee, 23.3% planted tea, 13.3% oranges, and 33.3% other economic crops.

S5 Rent out Farmland

Since a few years transferring user rights is allowed. About 21 out of 60 (35%) households have or had leased out land. Some consider it a source of income; other households do not have enough workers to cultivate the land themselves.

The policy on legalizing contracts for transfer of land use rights has offered chances to merge smaller pieces of land. In the area of the survey, several agricultural companies have also leased farmlands from farmers, and invested in constructing small reservoirs and spraying equipment for bananas. Although new legislation has made transferring user rights possible, the activity is not actively promoted through a government policy.

S6 Change Planting Schedule

TV and mobile phones are very popular in the rural areas of Yunnan province. Hence, the government can provide weather and drought information through TV stations and mobile phone companies and help local meteorological departments to issue weather information and warning messages.

However, the SMS services are not free. According to our survey, six households have changed their seeding time according to weather information, meaning that most of the households have not used the information.

S7 Adopt Improved Seed

This activity is very popular in the surveyed area. But this policy does little to improve farmers' adaptive ability to drought resistance. About 36 out of 60 (60%) households have taken improved (high yielding) seeds, and 12 out of 60 (20%) households have taken drought-resistant seeds. According to our survey, farmers prefer high quality and anti-pest seeds instead of high yielding or drought-resistant seeds of rice and maize, because in this area most of the grain is consumed by the farmers themselves.

The information on improved seeds usually comes from seed companies and community members. Farmers are stimulated by subsidies to buy improved seeds.

S8 Change Drinking Water Supply Mode

Drinking water supply is a public service offered directly by government, who is the only investor and heavily subsidizes this service. The farmers sometime need to provide some labor for obtaining this service. Of 60 households, 18 (19%) have taken measures to resolve drinking water shortages by connecting to the government supply system. This increases water supply and improves the possibility of water storage.

Sometimes local governments do not have enough funds to supply subsidized water. In that case the farmers depend on their own sources of drinking water supply.

S9 Construct Water Storage Infrastructure and Use Pumps

Small tanks or ponds for drinking water and irrigation are usually constructed by farmers themselves. About 17% of households own a small tank for drinking, and sometimes one for irrigation purposes. Big reservoirs or dams, which offer one or several communities irrigation water, require public funds. For 18 out of 60 plots on which we have information farmers used a pump to draw water from a channel.

Related policies of the government that were in the past based on the agricultural tax (*tongcou* and *tiliu*) have been canceled, and public funds for communities have shrunk. According to our survey, the policy to "raise money and labors case by case in the rural areas" (*yishi yiyi*) has been found to play an important role to compensate the shortage of public funds on the condition that the farmers come to an agreement on investing their labor and money in public goods. Local governments have offered money for reservoir construction and cement for dam construction in the area of our survey. However, only in one of the studied communities farmers have constructed their own small reservoir for irrigation, while two communities have received funds from their local government to construct a small dam for irrigation. With the expanding subsidy for agricultural equipment an increasing number of farmers use pumps. This has improved the farmers' ability to gain access to water resources. All farmers own one or more means of transport and at least one agricultural machine. The agricultural machine company mediates between the government and farmers. It has popularized the use of these machines in the rural areas.

S10 Change Irrigation Method

Increasing droughts and deforestation means that there is not enough spring water for irrigation, which is a traditional source of water. Farmers have to resort to new sources of irrigation, such as using canals and ground water.

Some policies have been implemented to alleviate their costs. Villagers can borrow pumps from local government, which uses its special funds for drought relief for this purpose. One of the villages studied has received diesel oil for pumping. Community leaders play an important role in mobilizing these resources from local government.

S11 Water Transfer
To resolve drinking and irrigation water difficulties, three villages have constructed new pipelines to transfer water from other mountains. Community members fetch water from another place alone or in cooperation when they face drought. Sometimes water is not free, because farmers need to pay local people. This could be considered the beginning of a water market. In other villages farmers use pipelines provided by their local government to import water if this is possible in the current situation.

S12 Dig a Well
In addition to investing as a single farmer, tube wells for irrigation can also be paid for by several farmers jointly. One household has constructed their own tube well mainly for livestock purposes. In the same village, a big tube well has been constructed by the government in 2010, a very dry year.

Digging your own well is not directly related to a policy and depends to a large extent on the income of the farmer concerned. However, sometimes, if the well is used for public purposes incentives can be obtained from the administrative village, such as the use of equipment or cement in combination with efforts by the farmers themselves.

S13 Apply Water Saving Technology
Only three households have ever chosen water saving technologies. They used cement to construct irrigation canals to prevent infiltration, using plastic mulch to store water more efficiently, or adopting spray irrigation equipment. There is no evidence that these activities are related directly to water saving policies implemented in China. The initiatives are started by farmers themselves with the main purpose of improving their revenue.

S14 Food Storage
This is quite common, but has become even more important given the increasing number of droughts. Food storing practices are not related to a

specific government policy but are even more profitable if the local markets do not function well.

DISCUSSION

Besides local government initiatives, some informal and formal organizations have helped to organize adaptive activities under the pressure of climate change. The formal organizations include community organizations, companies, special associations, and markets, while the informal organizations include lineages, neighbors, and relatives. Under pressure of the impending climate change and concomitant water scarcity, communities have undertaken collective actions. The policy of offering public goods gives a positive incentive to farmers to sit together to formulate an agreement on raising money and labor locally to undertake specific drought-related activities. Other activities, like keeping livestock, rotation, changing land use type, and food storage, are usually adopted by single farmer spontaneously without direct supports of policy.

Local companies also start to participate in drought relief activities when they can benefit from offering services. Special planting associations are organized by administration or natural village members and financially supported by local government to offer training courses, to invest in machinery, to facilitate purchase of seed, higher production promotion, and so on. Given weak formal organizations, the informal organizations are also very important in the study area. Some informal organizations are gradually becoming formal organizations. Water trading happens in the dry season and may lead to water markets. Several farmers cooperated in digging wells (an informal activity) and develop water users associations (a formal activity).

We noted an increasing self-consciousness at different levels of governance in China, including the farmers and entrepreneurs in the rural areas who are faced with a major challenge for which the government has no adequate reaction. The drivers of change lead to new governance structures emerging from below, which respond to the priorities of these groups. The drivers mentioned have created the enabling environment for the farmers to take initiatives and build their own governance structures. This may eventually lead to a very different situation in the rural areas, where no longer governments would take all the initiatives.

We conclude that the farmer's reactions can be classified as passive or active and the active reactions can be classified in traditional, new, and

Table 1. Classification of Farmer's Reactions to Drought.

Type of Farmer's Reaction	Reaction	Examples
Passive	Wait for government	Instructions to plant drought-resistant crops
Active		
1. Traditional	Using existing governance structure	Village leaders mobilize support at the administrative village level
2. New	Developing new or emerging governance structures	Farmers work together to build water supply work
3. Individual	Individual reactions	Farmers choose alternative options: trees, cattle, etc.

individual reactions (Table 1). Farmers use their social capital in water user associations and local drought relief teams to advance their solutions.

The adaptation strategies identified can be classified in seven categories, according to their objective and are as follows:

(1) To reduce the dependency on agriculture, especially traditional agriculture, nonagricultural work and livestock culture have been used;
(2) To reduce dependency on irrigation water, farmers have adopted rice–maize rotation, adjusted the crop structure, and changed land use type;
(3) To improve the drought-resistant ability of crops, farmers have changed planting schedule, and adopted improved drought-resistant seeds;
(4) To improve reliability of water supply, farmers have adopted centralized drinking water supply system and facilitated construction of water storage infrastructure;
(5) To improve ability to access to water resource, farmers have dug wells, transferred water by pipeline or vehicle, pumped water from channels, etc.;
(6) To reduce the consumption of irrigation water, water saving technologies have been adopted; and
(7) To avoid food shortage caused by drought, farmers, for example, seldom sell their own grain.

CONCLUSIONS

In this contribution we studied how in the case of drought, where the government is not always coming forward, the farmers take initiatives to adapt to the new reality. Different adaptation strategies are distinguished and their relation to different government policies is analyzed. Many policies are not implemented at local level because of the weak local government and the lack of suitable local organizations to make the policies work. Local governments are good at dealing with emergency disasters, but find it more difficult to encourage farmers to stand by when they face drought. We observed that farmers find their own solutions and create their own governance structures to assure for example supply of additional water to their fields.

Many policies are not implemented at local level because of the weak local government and the lack of suitable organizations. However, local government can adopt a more comprehensive approach to drought management than the national government. Local governments are better in dealing with emergency disasters, then in enhancing adaptive behavior with as a result that farmers stand by while they face drought. Some policies encourage community and other grassroots organizations to become stronger, and to start to play a more important role in dealing with daily climate risks. In response to the multiple hierarchical organizations operating independently in the field leading to policy gaps and overlaps, the central and local governments are increasingly using leading committees and integrated policies. However, farmers now try to organize themselves more at the grassroots level to adapt to new situation and to learn to cope with drought.

The mainstream of drought policies still lies in improving government's capability of offering public goods directly, although there appears a shift to give nongovernmental actors more space, because the awareness of local actors, networks, and knowledge are playing an important role in coping with climate change and drought disaster. Changing governance structures have brought some problems in coping with drought. Traditional agencies such as the technology promotion agency and the drought relief team have shrunk and are hardly capable to implement policies. Households are the major water resource managing unit. Except to offering direct help to farmers, it is hard for government to organize effective collective adaptation among farmers without bureaucratic agencies. Private and informal organizations have played an important role in adaptation.

Shifts in water governance structures to more adaptive governance improved the local ability to deal with the consequences of drought, but as Ostrom (1990) pointed out, decentralized water resource governance system also has several flaws, which should be considered, which include (a) the water resource is easily controlled by leaders or elite without a democratic decision procedure at the community level, (b) local people are apt to stick to past successful experiences while they find it hard to adjust to a changing environment, and (c) the water utilization conflict, which is caused by water scarcity.

After analyzing the top-down policies of the government, the initiatives of the farmers have been classified and the role of the government in these initiatives has been assessed. The farmers have learned how to benefit from government programs, but also that they should not wait for government support. It is important to utilize the indigenous knowledge of the farmers and their capacity to deal with the issues.

The effectiveness of China's policies to cope with drought is limited by financial challenges, organizational challenges, participation challenges, and market mechanism challenges. China needs to put more emphasis on enhancing adaptive capacity of institutions; drought management; enhancing coordination between different departments and levels of government; decentralization in policy making; and involving farmers or other stakeholders in decision making.

It was found that local government can adopt a more comprehensive approach to drought management than the national government. Farmers however, develop their own strategies in a situation where the government seems to be less present at the lower administrative levels.

Finding financing for climate change activities is important. Alternative financing mechanisms, such as the Clean Development Mechanism and Tradable Carbon Emission Rights could be used. The objective would be to increase the resilience of the different ecosystems in the Lanchang river basin. This requires better management, strong formal (public) and informal institutions, public private partnerships, sharing of knowledge, and leadership. In China there is a stronger emphasis on technocratic solutions and geo-engineering solutions, but the use of market mechanisms is still at a nascent stage.

In the Lanchang river basin, different reactions to drought were observed. They are only partially initiated by higher levels of governance. A number of factors seem to influence the number of initiatives coming from below. We concluded the farmer's reactions can be classified as passive or active and the active reactions can be interpreted as showing agency, used to advance their own solutions, in a situation where the government is not considered to be effective. More research on the livelihood strategies is necessary to better understand what these strategies mean for the household income and hence for the survival chances of different households.

NOTES

1. Lower and upper refer to the classification of the river within China and not to the Chinese versus the Laos, Cambodia, and Vietnam part of the river.

2. The website "http://www.ynszxc.gov.cn/S1/default.shtml" offers statistical data and also records public events and makes announcements for villages in whole Yunnan province. It helped to get a picture of the farmer's strategies to improve their livelihood, including their drought strategy.

3. In China farmland as well as water resources belong to the community and farmers obtain land use rights from the county or district authorities or can transfer user rights through contracting with other community members.

REFERENCES

Agrawal, A., & Perrin, N. (2008). *Climate adaptation, local institutions, and rural livelihoods.* IFRI Working Paper # W08I-6.

Barrett, C. B., Reardon, T., & Webb, P. (2001). Nonfarm income diversification and household livelihood strategies in rural Africa: Concepts, dynamics and policy implications. *Food Policy*, *26*(4), 315–331.

Benioff, R., Sandra, G., & Jeffrey, L. (Eds.). (1996). *Vulnerability and adaptation assessment: An international guidebook.* Dordrecht: Kluwer Academic Publishers.

Calow, R. C., MacDonald, A. M., Nicol, A. L., Robins, N. S., & Kebede, S. (2002). *The struggle for water: Drought, water security and rural livelihoods.* Groundwater Systems and Water Quality Programme Commissioned Report CR/02/226N. British Geological Survey Technical Report. Wallingford, UK: BGS.

Chiotti, Q., Johnston, T., Smit, B., & Ebel, B. (1997). Agricultural response to climate change: A preliminary investigation of farm-level adaptation in Southern Alberta. In B. Ilbery, Q. Chiotti, & T. Richards (Eds.), *Agricultural restructuring and sustainability: A geographical perspective.* Wallingford, UK: CAB International.

Downing, T. E., Ringius, L., Hulme, M., & Waughray, D. (1997). Adapting to climate change in Africa. *Mitigation and Adaptation Strategies for Global Change*, *2*(1), 19–44.

Goodwin, B. K., & Smith, V. H. (1995). *The economics of crop insurance and disaster aid.* Washington, DC: The AEI Press.

IPCC (Intergovernmental Panel on Climate Change). (2007). *Climate change 2007: Impacts, adaptation and vulnerability.* Contribution of Working Group II to the Fourth Assessment Report of the IPCC. Cambridge University Press, Cambridge.

Keshavarz, M., Karami, E., & Kamgare-Haghighi, A. A. (2010). A typology of farmers' drought management. *American-Eurasian Journal of Agricultural & Environmental Sciences*, *7*(4), 415–426.

Li, G. Q. (2010). The thinking of constructing Yunnan rural social security system. *Journal of Yunnan Agricultural University*, *4*(3), 10–14.

Li, H., Gupta, J., & Van Dijk, M. P. (2013). China's drought strategies in rural areas along the Lancang. *Water Policy*, *15*, 1–18. doi:10.2166/wp.2012.050

Li, H., & van Dijk, M. P. (2012). Climate change and farmer response in rural China. *International Journal of Water*, *6*(3–4), 290–311.

Nelson, G. C., Rosegrant, M. W., Koo, J., Robertson, R., Sulser, T., Zhu, T., ... Lee, D. (2009). *Climate change: Impact on agriculture and costs of adaptation*. IFPRI Food Policy Report.

Ostrom, E. (1990). *Governing the commons*. Cambridge: Cambridge University Press.

Paavola, J. (2008). Livelihoods, vulnerability and adaptation to climate change in Morogoro, Tanzania. *Environmental Science & Policy*, *11*(7), 642–654.

Parry, M. L., Rosenzweig, C., & Iglesias, A. (1998). Agriculture. In J. F. Freenstra, I. Burton, J. B. Smith, & R. S. J. Tol (Eds.), *Handbook on methods for climate change impact assessment and adaptation strategies*. Amsterdam: UNEP.

Rogers, P., & Hall, A. W. (2003). *Effective water governance*. TEC Background Papers No. 7. Geneva: Global Water Partnership.

SCNPC (Standing Committee of National People's Congress). (2002). *Water law*. Retrieved from http://english.gov.cn/laws/2005-10/09/content_75313.htm. http://english.gov.cn/laws/2005-10/09/content_75313.htm

Selvaraju, R., Subbiah, A. R., Baas, S., & Juergens, I. (2006). *Livelihood adaptation to climate variability and change in drought-prone areas of Bangladesh*. FAO report.

Smit, B., & Skinner, M. W. (2002). Adaptation options in agriculture to climate change: A typology. *Mitigation and Adaptation Strategies for Global Change*, *7*(1), 85–114.

Smith, J. B., & Lenhart, S. S. (1996). Climate change adaptation policy options. *Climate Research*, *6*(2), 193–201.

Talbot, C. (2005). Performance management. In E. Ferlie, L. E. Lynn, & C. Pollitt (Eds.), *The Oxford handbook of public management*. Oxford: Oxford University Press.

Tsai, L. L. (2002). Cadres, temple, and lineage institutions, and governance in rural China. *The China Journal*, *48*, 1–27.

van Dijk, M. P. (Ed.). (2012). Shifts in urban water governance paradigms, *Introduction to a Special issue of the International Journal of Water on Water Governance*, 6(3–4), 137–344.

Wilhite, D. A. (Ed.). (2005). *Drought and water crises: Science, technology and management issues*. Boca Raton, FL: CRC Press.

Ye, T., Shi, P., Wang, J., Liu, L., Fan, Y., & Hu, J. (2012). China's drought disaster risk management: Perspective of severe droughts in 2009–2010. *International Journal of Disaster Risk Science*, *3*(2), 84–97.

Yunnan Government Publication (YPG). (2008). Kunming: Yunan Printing Office.

Zhang, Y. F. (2011). Research on new countryside construction planning in minority areas. *Journal of Yunnan Administration College*, *1*, 158–161.

Zhou, Y., Lü, X. X., Xu, J. C., Zhang, H., & Jiang, T. (Eds.). (2011). *China climate change impact report: Yunnan province*. Beijing: China Meteorological Press.

IS ENVIRONMENTAL UNCERTAINTY REDEFINING FISHING STRATEGIES? THE USE OF THE TRADITIONAL LUNAR CALENDAR TO ALLOCATE FISHING EFFORT IN ENDE, EASTERN INDONESIA

Victoria C. Ramenzoni

ABSTRACT

Purpose – *The study assesses the significance of environmental uncertainty and its effects on fishing strategies of small-scale fishermen in Ende, Flores, Indonesia. Periodic environmental cycles such as the moon phase can have important effects on fishing strategies by regulating the behavior of stocks and tides. Traditional lunar calendars are used by subsistence fishermen to decide* when *and* where *to go fishing. Environmental uncertainty, specifically unprecedented changes in*

Climate Change, Culture, and Economics: Anthropological Investigations
Research in Economic Anthropology, Volume 35, 177–211

ISSN: 0190-1281/doi:10.1108/S0190-128120150000035008

oceanographic and atmospheric conditions, is threatening the predictability of traditional systems of ecological knowledge.

Methodology/approach – *Methods included ethnographic and observational techniques. Interviews* (n = *58*) *and surveys* (n = *132*) *are qualitatively and quantitatively analyzed. A combination of standard statistical tests, multilevel models, and cluster analysis is applied to long-term repeated observations of fishing events* (n = *2,633*).

Findings – *Endenese fishermen emphasized the importance of the traditional lunar calendar to allocate their effort in interviews and surveys. This belief does not coincide with observed behavior. Contrary to expectations from the traditional calendar, the lowest probability of fishing happens in the intermediate phases, with fishing also occurring during the full moon. Differences between individuals play an important role in explaining variability in returns. Finally, based on the consideration of variability, three different fishing strategies are identified that suggest an effect of environmental uncertainty in effort regulation.*

Research implications – *The paper underlines the importance of studies of variability to identify behavioral flexibility and adaptation. Results emphasize the value of considering individual traits in the analysis of subsistence practices.*

Keywords: Fisheries; traditional ecological knowledge; environmental change; fishing effort; foraging

INTRODUCTION

Unprecedented climatic change endangers the continuity of fishing livelihoods around the world (Daw, Adger, Brown, & Badjeck, 2009). It is estimated that within the next decades, rises in sea surface temperature and acidification will critically affect the survival of corals, shellfish, and finfish species causing changes in migration, recruitment, and reproduction processes (Cheung et al., 2009; Stocker et al., 2013). Other changes in ocean circulation and currents, as well as a projected increase in the frequency of extreme events like cyclones and storms (TROFCCA, 2006), will introduce further challenges to local economies dependent on marine resources for subsistence.

Anthropological, economic, and behavioral research in small-scale societies has shown that local mechanisms of prediction are central in securing livelihoods and adaptation (Godoy et al., 2009; Orlove, Chiang, & Cane, 2002; Tucker, 2007). For example, to guide decisions about effort and resource procurement, fishermen rely on a combination of cognitive skills, foraging techniques, and ecological cues (Bjarnason & Thorlindsson, 1993; Cordell, 1974; Morrill, 1967; Paolisso, 2002). Lunar calendars capture tidal patterns and other ecological factors affecting returns and are used by subsistence and recreational fishermen to decide *when* and *where* to go fishing.

New levels of environmental uncertainty and nonpatterned variability are testing the efficacy of traditional systems of ecological knowledge in terms of predictability (Green, Billy, & Tapim, 2010; Hamilton & Walter, 1999; Lefale, 2010). Changes in ecosystems and socioecological conditions have affected how fishermen perceive the structure of the environment and the value of seasonal fishing calendars (Ramenzoni, 2013). This, in turn, may have introduced the need for new strategies to organize subsistence activities.

Uncertainty can affect the state of natural resources by modifying levels of exploitation, influencing equipment and technology use, and increasing the likelihood of nonsustainable practices (Chakravarty-Kaul, 1998; Halim, 2002; Lowe, 2006; Mous et al., 2005; Pet-Soede, Cesar, & Pet, 2000). For example, as a consequence of high variability in catches fishermen may increase their use of already intensively harvested species by fishing for longer periods of time, expand to other fishing grounds thereby placing greater demand on stocks, rely on destructive fishing strategies to increase yields and/or contribute to the extinction of threatened groups by nonadhering to governance regulations. Because the application of policy requires in some cases a change in behavior, information on how people make decisions in terms of resource harvesting is central to eliminate errors in the design and implementation of governance measures (Aburto, Thiel, & Stotz, 2009; Branch et al., 2006, Defeo & Castilla, 2005; Salas, Sumaila, & Pitcher, 2004; Sethi, Costello, Fisher, Hanemann, & Karp, 2005).

In the next sections, this paper explores changes in fishing practices and effort allocation among small-scale fishermen in Ende, Flores, Indonesia. With this purpose, it: (1) investigates the use of the traditional Endenese lunar calendar to regulate fishing activities; (2) determines whether lunar phases have a significant effect on fishing and catch rates; (3) identifies profiles in temporal effort allocation that can be of value to policy makers; and (4) discusses the role of environmental uncertainty in shaping fishing behavior.

TRADITIONAL FISHING CALENDARS AND THE LUNAR CYCLE

The exploration of the association between the lunar cycle and fishing returns has offered insight into how biological processes can affect subsistence decisions and determine temporal patterns of effort (Bos & Gumanao, 2012; Lopeztegui-Castillo, Pinar, & Baisre Álvarez, 2013; Pet-Soede, Van Densen, Hiddink, Kuyl, & Machiels, 2001; Poepoe, Bartram, & Friedlander, 2007; van Oostenbrugge, Densen, & Machiels, 2001). This information is central to design sustainable conservation policies that are attentive to the preferences and behaviors of resource users (Aburto et al., 2009; Lopes & Begossi, 2011; Stelzenmüller et al., 2008; Teh et al., 2012). As some fish species tune reproduction and aggregation to moon periodicity and light availability, their behavior is highly predictable making them vulnerable to overexploitation (Poisson, Gaertner, Taquet, Durbec, & Bigelow, 2010). Knowledge of temporal patterns of fishing intensity can aid in the restoration of stocks by setting adequate closure dates.

Unfortunately, findings on the effect of the lunar cycle have been largely contradictory and vary based upon the species and the gear used (Poisson et al., 2010; Srisurichan, Caputi, & Cross, 2005; Stevenson & Millar, 2013; van Oostenbrugge, Bakker, van Densen, Machiels, & van Zwieten, 2002). In some case studies, results show substantial changes in returns during the full moon that can be several orders of magnitude (Bos & Gumanao, 2012; Di Natale & Mangano, 1995; Pet et al., 1997).

Likewise, ethnographic accounts from Australia, Southeast Asia, Polynesia, Brazil, and New England have detailed the extent to which fishermen rely on cyclical regularities such as the tides and the moon phases to allocate effort with traditional fishing calendars (Acciaioli, 2004; Acheson, 1981; Ammarell, 2002a, 2002b; Cordell, 1974; Malinowski, 1978; Millar, McKenzie, Bell, & Tierney, 1997; Poepoe et al., 2007; Ramenzoni, 2013). This effect was described by Malinowski in 1918 in the Trobriand Islands (1918, 1921, 1978) and also referred to by Firth in Malaysia in the early 1960s (1965 [2006]). A relatively well-known study from Cordell (1974) detailed the extent to which traditional fishermen in Brazil rely on cyclical regularities of the tides and the moon to decide when and where to fish each day. He emphasized that tides are one of the most important ecological factors because they determine fish behavior and the positioning of gear, thus affecting fishing techniques and yields. The moon has a critical influence on tides and currents through the gravitational pull. It also plays a role in explaining choice of fishing practices as it directly limits the duration and availability of light at night affecting the feeding behavior of fish (diel pattern). Cordell (1974) also suggested that fishermen's traditional knowledge of the tidal-lunar cycle reflected an adaptive strategy to procure a greater catch. In this way he assumed that the calendar was a reflection of an ecological reality: the reproductive and migratory behaviors of fish respond to moonlight and tidal conditions. Cordell contended that fishermen developed a system that could trace and predict through repeated experiences and memory (1974, p. 388).

In stark contrast to these accounts, other studies have not reported any significant changes in the impact of the lunar phase in returns; or have detected negligible influences (Lopeztegui-Castillo et al., 2013; Millar et al., 1997; Ortega-García, Ponce-Diaz, O'Hara, & Merilä, 2008; Stevenson & Millar, 2013). These results go against long-standing convictions maintained by traditional and recreational fishermen, biologists, and fishery managers about the abundance of certain species in the plenilune.

Factors that hinder the study of the moon cycle and its effect on catches include difficulties in the access to reliable datasets, the lack of "stringent statistical analyses" (Ortega-García et al., 2008, p. 198), and analytical limitations in dealing with the complexity of decision processes in fishery settings. These constraints are particularly important in the case of small-scale subsistence fisheries. Within artisanal or subsistence fisheries, longitudinal or panel research is infrequent mostly due to logistics and costs. Given the lack of temporal datasets with a high resolution, the identification of fishing allocation strategies is difficult. (Demanèche et al., 2008; Guyader et al., 2013; Pet-Soede et al., 2000; Salas et al., 2004; Zeller, Booth, Davis, & Pauly, 2007; Zeller, Booth, & Pauly, 2006).

Probably one the most significant methodological limitation arises from the way that statistical analyses of subsistence behavior treat individual fishing events as independent data points (Lopes et al., 2011; Ruttan & Tyedmers, 2007; Stephens, 2008; Westneat & Fox, 2010; Winterhalder & Smith, 2000). For example, in Optimal Foraging Theory, researchers collect observations of fishing trips and estimates of catches within particular communities of fishermen (i.e., Aswani, 1998; Begossi, 1992; Lopes & Begossi, 2011; Silvano & Begossi, 2001; Sosis, 2001, 2002). To analyze the data, each single fishing event is aggregated to a dataset, which is then summarized in means and rates. Individual characteristics of the fishers are frequently used to weight the value of observations and are often corrected for in the analysis. In the end, Optimal Foraging Theory (OFT) explanations refer only to the average generic person or group (for an exception, see Koster, Grote, & Winterhalder, 2013; McElreath & Koster, 2014).

Because in fishing effort studies participants are repeatedly sampled over consecutive days, months, and seasons, data points show correlations that violate the assumptions of independence of observations presupposed in the statistical analysis. Correlations are due to individual elements that might bias measurements (fishing skills), group effects (age), or to the closeness of measurements in time. They can be responsible for some of the variation that remains unaccounted by the explanatory variable. Most significantly, if individual traits are not included in the analysis they can lead to biased results and incorrect conclusions (Hayes, 2006; Salas & Charles, 2007, p. 118; Snijders, 2011; Snijders & Bosker, 1999).

In the last 10 years, mixed models, also known as multilevel or hierarchical models, have become more common in economics, behavioral studies, and biostatistics. They permit the simultaneous exploration of the different kinds of variability in the measurements by considering random and fixed effects and by including multiple error terms (Bolger, Davis, & Rafaeli, 2003; Bolger & Laurenceau, 2013; Cnaan, Laird, & Slasor, 1997, p. 2349; Ott & Longnecker, 2010). This gives them more flexibility and makes them a good alternative in datasets where individual profiles are of interest. Complex approaches to fishing patterns have looked at cluster associations (Mahévas, Bellanger, & Trenkel, 2008; Tzanatos, Somarakis, Tserpes, & Koutsikopoulos, 2006), principal component analysis and factorial techniques (Hamon, 2009; Pelletier & Ferraris, 2000), dynamic studies of fleets (Branch et al., 2006), time series analysis (Marchal, Poos, & Quirijns, 2007; Srisurichan et al., 2005; van Zwieten, Roest, Machiels, & Van Densen, 2002), and data envelopment analysis (Vázquez-Rowe & Tyedmers, 2013). However, it is only very recently that scholars are applying multilevel analysis to the consideration of fishermen's behavior (Álvarez & Schmidt, 2006; Millar et al., 1997; Millar & Willis, 1999; Ruttan & Tyedmers, 2007; Stevenson & Millar, 2013). Furthermore, despite the large number of studies on fishing patterns, with the exclusion of a few cases (Pet-Soede et al., 1999, 2001; Salas & Gaertner, 2004; Salas et al., 2004; van Oostenbrugge, 2001), fishery research on individual decision processes has shown a preference to the study of the spatial dimension of effort allocation to the detriment of the temporal dimension (van Densen, 2001, p. 13). Temporal studies require consistent and prolonged data collection of individual practices, environmental variables, and other contextual events that might influence effort to permit the detection of decision patterns that reflect changes in conditions over time. Because such datasets, especially in small-scale subsistence fisheries (Zeller et al., 2006, 2007), are not common, there is a need for a more detailed analysis of temporal dimensions to better understand extraction practices (van Densen, 2001). This paper directly addresses this gap by centering on the study of repeated choices through a multilevel statistical approach.

A second methodological reason why findings might contradict expectations of the role of the lunar cycle is related to the difficulty in accounting for the complexity of the decision process. Subsistence fishermen are a highly heterogeneous group in terms of the technological characteristics of the gear they use and of their preferences (van Oostenbrugge et al., 2001; Wiyono, Yamada, Tanaka, Arimoto, & Kitakado, 2006). At any particular time, a fishery exhibits multiple different arrangements of resource exploitation, with individuals fishing independently, in small corporate groups or even in associations. These arrangements are difficult to characterize in totality and in a consistent fashion. As a result, researchers tend to aggregate information of individual decisions to the group or community level, and lose resolution (Salas & Charles, 2007).

Moreover, variability in decision-making often results from individual preferences, characteristics, and skills that might not be explained by technical or ecological factors alone (King, 2011; Ramenzoni, 2013; Salas & Charles, 2007). This has led researchers to hypothesize the existence of a "skipper effect" which builds on the belief that skippers play a determinant role in fishing success due to personal attributes such as individual luck and skills (Acheson, 1981; Durrenberger & Pálsson, 1986; García-Quijano, 2006; Ruttan & Tyedmers, 2007; Thorlindsson, 1988). The relevance of individual traits in explaining success can make the generation of models or profiles difficult as the motivation behind strategies is not always the same (Álvarez & Schmidt, 2006, Ruttan & Tyedmers, 2007).

Finally, another reason affecting the study of the relations between lunar cycles and the regulation of fishing effort may be found in the complex set of modifications in oceanographic and atmospheric conditions associated with the joint effects of environmental change and overfishing (Bell et al., 2013; Daw et al., 2009; Roessig, Woodley, Cech, & Hansen, 2004; Stocker et al., 2013). Some of these modifications, such as the reduction of net primary productivity and biomass or changes in the length and intensity of monsoon seasons in the Indo-Pacific Ocean, have significant consequences for both fish and fishermen. There is evidence that, in order to respond to overfishing and ecological changes, fishing effort patterns are changing (Green et al., 2010; Lefale, 2010). For example, variations in oceanographic conditions – such as changes in circulation, tidal regimes, and currents – as well as an increase in the frequency of extreme events have also presented new challenges to fishermen by constraining accessibility to fishing grounds and navigation, and by increasing the risks in operations (Stocker et al., 2013; TROFCCA, 2006). Given the velocity and uncertainty surrounding environmental change, highly dynamic conditions may underscore the inadequacy of previous fishing strategies such as the lunar and seasonal calendars, and suggest the need for rapid adaptation. However, this idea remains unexplored.

PREDICTIONS

In this paper, I rely both on classic statistical techniques and mixed methods to investigate whether fishermen in Ende, Flores, Indonesia, allocate their temporal effort according to the lunar cycle. By exploring the existence of changes in the use of the calendar to organize fishing practices, I examine the effect of environmental uncertainty in behavior regulation. It is assumed that if moon cycles have a demonstrable effect on catches, then Endenese fishermen would not go fishing on unfavorable days (during and surrounding the full moon) or would catch less fish as represented in catch rates. The following predictions are tested:

1. Lunar phases affect probability of fishing with lower probabilities of fishing on a given day occurring during the full moon in relation to other phases of the month (half-moon and new moon).
2. Lunar phases affect catch rates (catch per unit of effort or CPUE), with lower CPUEs per subject during the full moon in relation to other phases of the month (half-moon and new moon).
3. Fishermen that fish close to the full moon more frequently would have higher variability in their catches in comparison to fishermen that fish more often in the other two phases (half-moon and new moon).
4. Fishermen that fish close to the full moon more frequently will have higher variability in fishing time in comparison to fishermen that fish more often in the other two phases (half-moon and new moon).

   To discuss the role of environmental uncertainty in fishing behavior, I assume that environmental uncertainty will reduce or modify effort by affecting navigation and accessibility to fishing grounds. If this expectation is correct, I anticipate that to secure returns fishermen may go fishing when conditions are adequate in spite of the lunar cycle. Prediction 5 states that:

5. Changes in fishing preferences may translate into new behavioral strategies that show a high tolerance for variability in daily returns and time spent fishing.

ETHNOGRAPHIC BACKGROUND

People and Economic Activities

Ende (W 121.519460 and S 8.879934) is a small port located in the Bay of Ende and the capital of the Regency of Ende (Kabupaten Ende), central Flores, Nusa Tenggara Timur, Indonesia (Fig. 1). Across the bay from the city is Pulau Ende, an 8 by 2 km island, with a population of approximately 8,000 souls distributed in 7 villages. There are about 1,700 fishermen in Pulau Ende and about 2,000 in Ende city (Badan Pusat Statistik Kabupaten Ende, 2012). Endenese are a mix between hinterland groups and migrant Makassarese, Buginese, Javanese, and Chinese traders (Dietrich, 1983; Nakagawa, 1984; Ramenzoni, 2013). Throughout the nineteenth century, Endenese were known as powerful seafarers and pirates (Needham, 1968, 1980), engaging in slave trading and commercial activities with partners in modern Sulawesi, Timor, and Singapore. After Dutch military colonization, taxation, and forced labor,

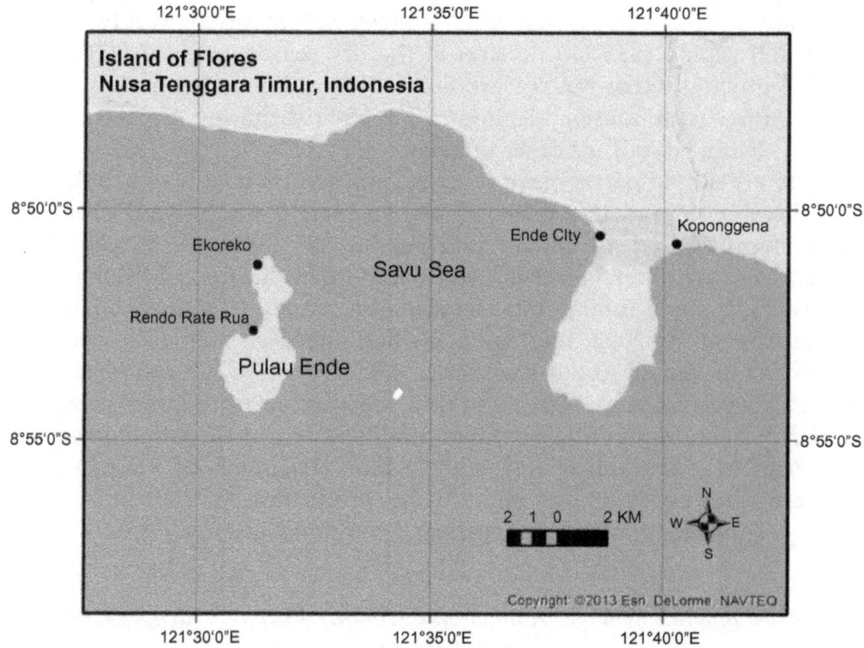

Fig. 1. Map of Ende and Pulau Ende, Island of Flores, Eastern Indonesia.

in the early twentieth century coastal Endenese sedentarized and transitioned to subsistence-based fishing and agricultural economies.

Nowadays, fishing activities are for household consumption and to support small-scale trade at the local market. The list of harvested species comprises *Scombridae*, *Clupeidae*, *Lutjanidae*, *Serranidae*, and *Carangidae* families. Typical fishing equipment includes gillnets (1–4.5 in. mesh, 5–7 pieces) and hand-line fishing and delineate two types of fishing métiers. A métier can be defined as a group of fishing activities that are characterized by the gear, the type of species targeted, the area of operations, and the season (Tzanatos et al., 2006). In Ende, the first and most popular kind of métier includes hand-line fishing with the use of spears and butterfly nets over reef barriers or shoreline areas (less than 2 miles) with canoes (100–200 kg). The second most common métier includes the use of gillnets, which are deployed in relatively deep areas (depths of about 100 to 1,000 m, more than 5 miles offshore), and require the use of motorboats (>500 kg). In both cases, while at sea, fishermen also make extensive use of equipment such as lures and lamps. Fishing trips begin around 5 pm and last on average 6 hours, but can extend to over 10 hours. Basic primary information from surveys indicates a mean self-reported weekly income of 380,000 rupiah ($35 US dollars) in the dry season and 125,000 rupiah ($10 US dollars) in the wet season. Sharing constitutes a significant subsistence strategy with households reporting the redistribution of parts of their yields with a mean of four other families.

There are no industries based in Ende, nor private investment to develop the fishery. Pole and line fishing fleets operate in the Savu, Timor, and Indian oceans based in Kupang and Benoa, Bali (Stacey et al., 2011). In the last 25 years, the region has experienced intensification of fishing activities. In 1986, Ende fishing fleet included 196 canoes, 613 plank boats, and 72 motorboats. By 2012, these numbers had increased manyfold, while the CPUE has decreased about four times. Oscillations in the composition of the yields since the late 1980s point to a substitution from larger pelagic to smaller pelagic catches and potential modifications in the trophic levels of some species. The studied area had an annual landing of 918,640 kg in 2010 constituting 12.8% of the marine production of the entire Ende regency (Dinas Perikanan Ende, personal communication).

Environmental Change, Annual and Monthly Fishing Calendars, and Local Perceptions

The marine ecosystem of the Savu Sea, where Ende is located, can be characterized as tropical, with sea surface temperatures approximating 28°C throughout the whole year. Primary productivity is about 561 $mgC \cdot m^{-2} \cdot day^{-1}$ and is significantly affected by the action of the monsoons (Potemra, Hautala, & Sprintall, 2003). As a consequence, environmental patterns have an important effect in the timing and duration of fishing seasons (Pet-Soede et al., 2001; Wiyono et al., 2006).

Since the early 1980s, important environmental changes seem to have occurred in Eastern Indonesia with a higher frequency of El Niño/La Niña Southern Oscillation (ENSO) events in the Indo-Pacific area (Annamalai & Liu, 2005; Pasaribu, 2007; TROFCCA, 2006). During El Niño periods, the surface easterly winds are interrupted or weakened, and the thermocline in the Eastern Pacific Ocean deepens creating abnormal sea surface temperatures in the area (Dayem, Noone, & Molnar, 2007). As a result, there is a shift in the pressure zones further modifying precipitation regimes, especially in the Eastern part of Indonesia and during the dry season (Hendon, 2003, p. 1782; van der Kaars et al., 2010). With longer drought periods and other changes in winds, rains, and ocean circulation that can be associated with more extreme ENSOs, the duration of the monsoon seasons might have changed affecting the predictability of climatic events. Environmental cues have become highly indeterminate in their capacity to anticipate the onset and duration of wet seasons which, in turn affects yields and fishing calendars (Ramenzoni, 2013; Dinas Pertanian Ende, personal communication).

For example, in annual fishing calendars, the allocation of fishing effort reflects the migratory and reproductive behaviors of fish as they occurred in certain months of the year given climatic and oceanographic conditions. Knowing that the fishing season for small scad (*Atulae or Decapterus* sp.) happens in January, Endenese fishermen would strategize and parcel their activities in the previous months to prepare for that event. But, as indicated for other species (Lehodey, Bertignac, Hampton, Lewis & Picaut, 1997; Lehodey, Chai, & Hampton, 2003) and mentioned in interviews and numerous informal conversations, recruitment processes, and migratory patterns have been altered by climatic variability and overfishing (Ramenzoni, 2013).

With meteorological seasons and annual fishing calendars presenting higher uncertainty, fishermen have begun to target multiple species by diversifying fishing tools, by incorporating innovations such as modern baits, and, most significantly, by altering their pattern of activities in the wet season (Ramenzoni, 2013). Prior to adopting these innovations, fishermen would remain at home for a period of 40–50 days (in December, January, and February) as strong westerly and northern winds prevented navigation. At present time (year 2012), the duration of the storms seems to have changed as precipitation events have shortened and become more fragmented (Dinas Pertanian Ende, personal communication). This allows fishermen to go fishing throughout the year, remaining inactive occasionally for periods of one or two weeks when storms and cyclones hit the region. As mentioned in interviews and conversations, the frequency of their trips in the wet season has changed (Ramenzoni, 2013).

Conversations with fishermen also pointed to the significance of the moon cycle and their influence on tides and fish to regulate behavior (Ramenzoni, 2013). However, fishermen also mentioned important changes in sailing conditions, ocean currents, and tidal regimes affecting their daily and monthly effort allocation. In the next sections I analyze interview and observational data from fishing diaries to explore the role of the moon (predictions 1 through 4) and environmental uncertainty (prediction 5) in shaping fishing strategies.

METHODS, DATA, AND ANALYSIS TECHNIQUES

Methods

Ethnographic data collection (58 semistructured interviews and 132 surveys) and participant observation were conducted during 22 months of fieldwork from May until August 2009, from November 2010 until January 2011, and from June 2011 until January 2013 in Pulau Ende and Ende city, Flores, Indonesia. To monitor the allocation of time to fishing activities, surveys, observations, and fishing diaries were utilized at the villages of Ekoreko and Rendo Rate Rua in Pulau Ende from September 2011 until July 2012 (9 months). Fifty fishermen households were recruited to be a part of the longitudinal project by completing diaries (30 subjects) or by taking part in daily questionnaires about effort and focal observations (20) that spanned 10 months of research. Participants were drawn from a larger sampled population of 130 individuals with mean age of 41.8 years (SD = 12.9). Of these initial participants, only 25 fishermen (12 motorboat fishers and 13 canoe fishers) had sufficiently complete records for the duration of the study. Given differences in individual time allocation, initial and final dates of inclusion in the study varied depending on the fisherman. A total of 2,633 observations were recorded, including hour of departure, return, type of boat and gear, species and total number of fish caught, and the reason for not going fishing.

Data Description and Measures

Traditional Lunar Calendar, Lunar Phase, and Seasonality

To reconstruct the traditional lunar calendar, 58 fishermen were asked in semistructured interviews to provide rough estimations of the amount of fish that can be expected to be caught during the lunar month (e.g., on first day, during the full moon, day 22nd, day 30th). In addition, 132 fishermen participated in a comprehensive survey that included questions about general fishing activities, preferences, and effort regulation.

To study actual effort allocation, days of observations from fishing diaries were numbered consecutively for each participant depending on the day of the year (annual date). Then, observations were matched against the lunar cycle by looking at the US Naval Observatory calendar and assigned to the corresponding lunar day and lunar phase. Two variables were derived: (1) Lunar day is a continuous ordinal variable that redescribes the current date on a 29–30 day moon cycle; and (2) Lunar phase is a categorical ordinal variable that indicates whether a particular day belongs to one of three classes (full moon, half-moon, and new moon) according to the position of the moon (Pet, Van Densen, Machiels, Setyohadi, & Tumuljadi, 1997; van Oostenbrugge, 2001; van Oostenbrugge et al., 2001; Fig. 2). The following criteria were used to construct the classes:

- The full moon class begins from the 11th day on a lunar moon and lasts until the 21st. It includes days of the full moon and the waxing and waning gibbous. Because fishermen rely on lamps and bait to catch fish, this is the time when catches are the lowest and approaching zero.

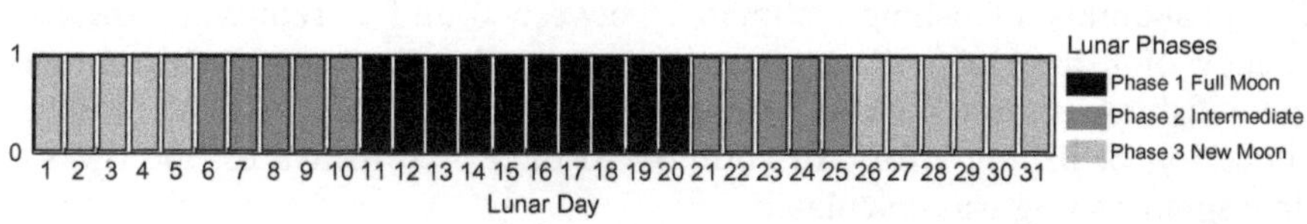

Fig. 2. Categorization of Lunar Days in Phases. *Notes*: The figure represents the division of the lunar days according to three phases that were used to analyze the dataset. Phase 1 or Full Moon comprising days 11–20, Phase 2 or Intermediate comprising days 6–10 and 21–25, and Phase 3 or New Moon comprising days 1–5 and 26–31 of the lunar month.

- The half-moon class comprises the days between the 21st and the 26th and the 6th and the 11th. This is when the moon is in its first and third quarters (waxing and waning gibbous).
- The new moon class includes days from the 26th until the 6th. This is when the moon is waning crescent or waxing crescent. Usually, the new moon day is not considered a good day for fishing.

A third variable was constructed to measure the proximity of the observation to the full moon. The variable was calculated by creating weights reflecting the distance of the event in terms of the moon date to the plenilune. The weights included values from 1 to 0.10 of a point. For example, if the day in the lunar calendar was 14, 15, or 16, a value of 0.1 was assigned, if the day of the observation was during the half-moon, values oscillated between 0.2 and 0.5.

Catch

In order to measure the success of fishing trips three variables, obtained from the fishing logs and observations, were explored: total catch (number of fish caught on a particular day), catch rate or CPUE (number of fish caught divided by the time spent fishing measured in minutes), and coefficient of variation of catches (CV, standard deviation of total catches divided by mean catches). Because fishermen did not report cases when the catch was zero, CPUEs and CVs were used to avoid bias in the analysis given the high skewness of the data. To control for the role of individual traits in testing predictions 2, 3, and 4, CPUEs were normalized through the use of *Z*-scores and CVs were also included in the analysis.

Time Allocation to Fishing and Effort

The probability of fishing (estimated between 0 and 1, represents the frequency of fishing trips in a given lunar calendar day) and total minutes spent fishing (minutes spent fishing by individual) were used to monitor allocation of fishing effort. In addition to these two measures, the CV for time spent fishing was calculated.

Missing Values and Zero Data Events

The lack of an entry on a diary or a blank observation was considered a missing value if ethnographic, anecdotal, and meteorological information did not suggest adverse conditions for fishing. To deal with missing data, a two-step analysis was conducted: one that included missing values and one that did not. Observing no significant differences, analyses included missing values when possible (Bolger & Laurenceau, 2013).

Data Analyses

Responses from semistructured interviews about estimations of catch in the lunar month were first analyzed qualitatively to establish consistency in estimations and then quantitatively. Surveys, diaries, and observations were analyzed quantitatively. All statistical analyses where conducted with the software JMP Pro Version 10.0.1 Release 2 from SAS. The choice of techniques was in accordance to the hypotheses tested. Regressions and tests accounted for the type of fishing vessel or métier (canoe and motor boat) by fitting different models or by including a variable denoting the kind of boat when appropriate. In addition, initial exploratory analyses (time series and nonparametric association tests; Millar et al., 1997; Poisson et al., 2010) showed an effect of seasonality on total catches indicating the need to control for this variable in some of the models.

Prediction 1: Binomial Logistic Model

According to previous research, to test hypothesis 1 the association between lunar phase (full moon, half-moon, and new moon) and probability of fishing (0 to 1) was modeled for each type of métier with a binomial logistic regression (Pet et al., 1997; Pet-Soede et al., 2001; van Oostenbrugge et al., 2001). A logit link function was used given that lunar phase displays a sinusoidal shape (log distribution). The following model was constructed for the linear predictor $g(m)_i$:

$$g(m)_i = \ln \frac{p(c)}{1 - p(c)} = \beta_0 + \beta_1 l_i$$

where β_0 and β_1 are unknown coefficients, and l_i is the effect of the lunar phase.

Prediction 2: Multilevel Model

First, an Interclass Correlation analysis tested the level of autocorrelation of observations. The test allows for the determination of how much variation in a longitudinal panel study can be attributed to the effect of individual differences (between-subjects) or to external differences (within-subjects) in the observations. Between-subjects effects are, in this case, individual traits that can influence catch rates including skills, personality characteristics, fishing preferences such as choice of gear and spots (Millar et al., 1997). Because this particular dataset includes repeated measures of fishing events carried out by the same fisherman over time, it is expected that CPUEs will reflect between-subject effects (Álvarez & Schmidt, 2006; Ruttan & Tyedmers, 2007; Stevenson & Millar, 2013).

Second, to address the potential existence of individual effects and the intercorrelation of observations, a mixed effect multilevel analysis was proposed. The Individual identification (ID) in each observation, which marked the fisher identity, was used as the random effect variable. Changes in CPUEs in terms of lunar phases (prediction 2) were modeled according to the following expression:

$$Y_{ij} = \gamma_{00} + \gamma_{01}E_j + \gamma_{10}L_{ij} + \gamma_{11}E_jL_{ij} + u_{0j} + u_{1j}L_{ij} + \varepsilon_{ij}$$

where Y_{ij} is the CPUE for subject j on a particular time point within a certain lunar phase; γ_{00} is the common (fixed) intercept for all subjects, $\gamma_{01}E_j$ the common (fixed) effect of vessel type or métier (dual level) for population γ_{01}, $\gamma_{10}L_{ij}$ the (fixed) effect of lunar phase for population γ_{10}, $\gamma_{11}E_jL_{ij}$ the (fixed) interaction of vessel type by lunar phase, u_{0j} subject-specific intercept deviation (random effect), $u_{1j}L_{ij}$ subject-specific slope deviation by lunar phase (random effect), and ε_{ij} is a residual within-subject error term.

Predictions 2, 3, and 4: CV and Z-Scores of CPUEs

To further consider the role of individual traits in the associations between catches, time fishing, and the lunar cycle (predictions 2, 3, and 4), normalized CPUEs, CVs, proximity to the full moon, moon phase, and type of vessel were used. The correlations among these variables were studied through nonparametric tests (Kruskal–Wallis and Wilcoxon) and regression.

Prediction 5: Cluster Analysis

Cluster analysis was used to further understand if patterns existed in the way fishermen allocated their time considering variability in catches and proximity to the full moon. This allowed the exploration of prediction 5. The way the analysis works relies on the grouping of a set of observations according to their similarity. The level of similarity is based on the consideration of several continuous responses. In this case, individual CVs of catch, fishing time, and average proximity to full moon were considered as the clustering variables. To test the significance of the clustering, further parametric tests of means (Wilcoxon) were calculated for each cluster. The groups are used later to propose profiles or strategies to deal with the variability in returns (Cordell, 1974; Salas et al., 2004).

RESULTS

Semistructured Interviews and Surveys

According to responses to semistructured interviews, fish abundance closely reflects the moon cycle, with catches increasing at the beginning of the lunar month, decreasing to values approaching zero during the days surrounding and comprising the full moon, and recovering after the 22nd day of the month. The final day of the calendar is considered a bad day for fishing given changes in tides (Fig. 3). The behavior of catches in the four weeks comprising the moon cycle is summarized in the following rule expressed by many fishermen during the interviews: "*ikan naik [minggu pertama], turun [semakin dekat bulan terang], naik [minggu ketiga], dan turun [minggu keempat]*"; which translates to: fish increase [in the first week of the cycle] and decrease [as the full moon approaches], increase [in the third

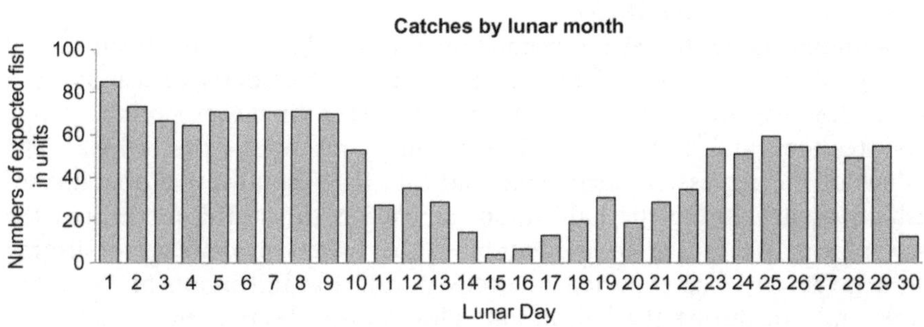

Fig. 3. Catches in Units of Fish According to the Lunar Month. *Notes*: The chart represents the verbal responses of 58 Endenese fishermen to the numbers of fish that can be caught during the lunar month. The bars represent the average values for each day. Of importance is the catch approaching zero values around day 15. A spike in catches is observed starting day 19 or day 23. Finally, catches approach zero again in day 30 as no fishing can be conducted in that day.

week], and decrease [in the last week of the month]. These four phases roughly overlap with the intermediate, full moon, second intermediate, and new moon phases used in the subsequent analyses of fishing diaries. Most importantly, the rule informs fishermen's expectations, organizes effort, and is often invoked to explain the lack of substantial yields in the plenilune. For example, fishermen would frequently mention that the dark period or "*bulan gelap*" was the best moment to catch fish. In addition, they would also indicate that "*bulan purnama*" or the full moon was not good as fish could see the nets and hooks and would not eat. Results from surveys to estimate fishing intensity indicated that 82% of participants fish between 15 and 25 days a month and largely avoid activities in the full moon. They also indicated that fishing expeditions extend for 6–7 hours reflecting the behavior of the tides. Only once during informal conversations and interviews, a fisherman mentioned going out on the full moon as prices of fish were higher during that time. Women were also aware of the cycle, timing their harvesting of octopus and shells to the neap tides, with harvesting occurring four to three days before the plenilune and after.

It must be observed that, even when there is a precise knowledge of the number of fish that can be expected to be caught on critical days of the cycle, of the behavior of tides and currents, and of the slight changes in the position of the moon each day, fishermen do not seem to have special terms to designate critical temporal marks in the cycle.

Prediction 1: Binomial Model

A binomial model tested the prediction that the lunar cycle affects actual fishing effort. The effect of the lunar phase on probability of fishing on a given day was significant ($p < 0.05$; Table 1) with the exception of one case (canoes and full moon). But results contradict expectations. The values of the regression coefficients indicate that the lowest probability of fishing occurs during the half-moon in relation to the full moon and the new moon. According to prediction 1, the lowest probability of fishing on a given day was expected to occur during the full moon of the lunar cycle and not during the half-moon. This is also observed in a simple frequency analysis of time spent fishing according to calendar day (Fig. 4). In addition, the regression models do not seem to be relevant in terms of how much variation is explained (Adjusted R^2, a parameter assessing model's fit, is less than 7%). This might be a consequence of the selection of the variables to explain probability of fishing on a given day and may reflect the importance of considering the effect of individual traits.

Table 1. Binomial Models.

Métier Canoes				Métier Canoes			
Linear model				Linear model			
Distribution: Binomial – Link: Logit				Distribution: Binomial – Link: Logit			
Source	DF	χ^2	Prob > χ^2	Source	DF	χ^2	Prob > χ^2
Lunar phase	2	18.54	<0.0001*	Lunar phase	2	23.47	<0.0001*
Parameters				Parameters			
Term	Estimate	χ^2	Prob > χ^2	Term	Estimate	χ^2	Prob > χ^2
New moon	0.175	10.08	0.0015*	New moon	−0.118	4.453	0.0348*
Full moon	0.003	0.002	0.96	Full moon	−0.173	4.777	0.0288*
Intermediate	−0.293	14.18	0.0002*	Intermediate	−0.207	6.850	0.0089*

*Indicates significance at α: 0.05.

Prediction 2: Multilevel Model

The Inter Class Correlation (ICC) analysis assessing the autocorrelations of measurements reported that about 23% of variability in the sample originates from a between-subjects effect (third class type of correlation) suggesting that observations are not independent and justifying the use of a multilevel model. The model tested prediction 2 with individual fishermen's IDs as a random component. The Adjusted R^2 captured 18% of the variability in the dataset (Table 2). Results suggest the following points: First, the estimated variation between fishermen is 21% of the total variation in the random effect component pointing to the importance of individual traits in results. Second, lunar phase is nonsignificant as a predictor for individual catch rates ($p = 0.61$). Third, vessel type was nonsignificant in explaining daily catch rates ($p = 0.37$). Fourth, the interaction between lunar phase and vessel type was highly significant ($p = 0.02$). One way to interpret these findings implies that CPUEs differ according to changes in the relation between lunar phases and type of vessel and in spite between-subjects variation. However, results also suggest that the way the CPUEs vary is not the same for all subjects and that this variation needs to be taken into account.

Predictions 2, 3, and 4: CV and Z-Scores of CPUEs

To further assess the role of the lunar phase in catches and effort and to explore differences by kind of métier, the *Z*-scores of CPUEs per subject during the full moon were matched against the intermediate phase and the

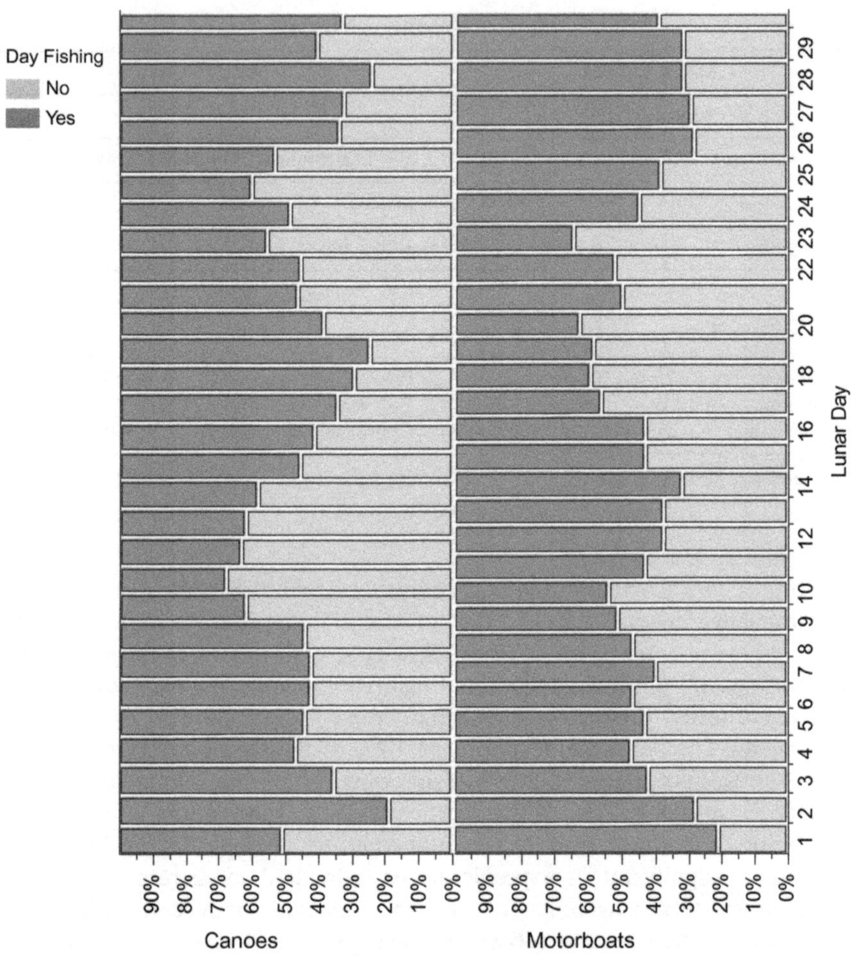

Fig. 4. Frequencies of Time Spent Fishing during the Lunar Month. *Notes*: The charts present the observed percentage of time spent fishing during the lunar month by day per métier. Observe the activity of both types of vessels during the lunar days 12 to 19. This is the time when the percentage of time allocation to fishing should approximate zero.

new moon values according to their vessel type. The comparison (Kruskal–Wallis Test) among normalized CPUEs in different phases of the moon only reported significant results for motorboats between the full moon and the new moon, and between the half-moon and the new moon.

Table 2. Multilevel Model.

Random effects parameters			
Covariance parameter	**Estimate**	**Standard error**	
ID	20.58	6.64	
Residual	70.02	2.68	
Fixed effects parameters			
Term	**Estimate**	***t* Ratio**	**Prob > |*t*|**
Intercept	6.19	6.55	<.0001*
Lunar phase (FM)	−0.32	−0.98	0.32
Lunar phase (I)	0.13	0.40	0.68
Motor (0)	−0.85	−0.91	0.37
Motor (0)×Lunar phase (FM)	−0.53	−1.63	0.10
Motor (0)×Lunar phase (I)	−0.28	−0.84	0.40
Fixed effects tests			
Source	***F* Ratio**	**Prob > *F***	
Lunar phase	0.49	0.61	
Motor	0.82	0.37	
Motor × Lunar phase	3.60	0.02*	

*Indicates significance at α: 0.05.

This is explained by the fact that CPUEs are significantly higher in the new moon, indicating that fishing in the full moon or in the half-moon phase may not be as profitable. In the case of canoes, no difference in means according to the moon phase was found.

When exploring the association between individual CVs in catches and type of vessel, a Wilcoxon test reported no significant effects ($p = 0.60$). Because CVs capture the amount of variability experienced by the subject, this finding indicates that relying on motorboats rather than canoes does not seem to make catches more stable. In short, results neither fully support prediction 2 nor indicate a difference in CVs by type of vessel or métier.

In relation to predictions 3 and 4 that explored CVs in the different phases of the moon, results show that CVs in catches significantly decreased as fishermen operated further away from the full moon ($p = 0.00$, Adjusted $R^2 = 0.30$; Fig. 5). CVs of time spent fishing also diminished progressively closer to the new moon ($p = 0.00$, Adjusted $R^2 = 0.62$; Fig. 5) as fishermen seem to allocate more time consistently during this phase.

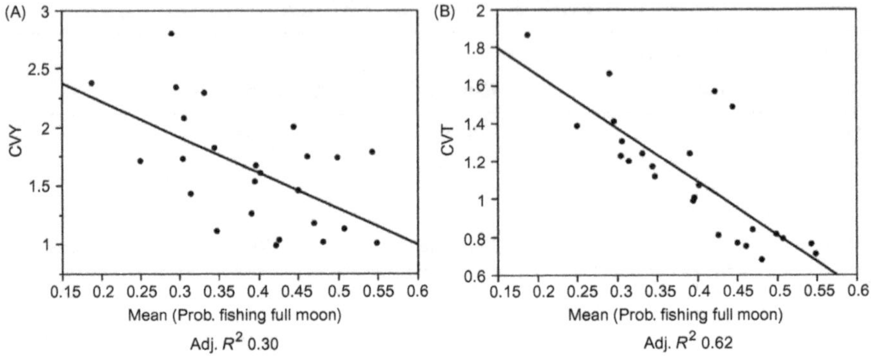

Fig. 5. Associations between CV and Time Fishing and Mean Probability of Fishing in the Full Moon. *Notes*: Part A shows the negative association between individual CVs in catches and probability of fishing close to the full moon for the 25 fishermen sampled. Part B also shows the inverse association between time spent fishing and the probability of fishing close to the full moon.

Prediction 5: Cluster Analysis

Results from the cluster analysis indicated the existence of three groups based on CV in catches, time spent fishing, and proximity of fishing to full moon. All parametric Wilcoxon tests reported p values of 0.001 or below denoting that clusters were significantly different among each other. The first cluster included six fishermen that have, on average, lower CVs in yield than the rest of the sample (Fig. 6). They also fish with more frequency during the new moon phase and have high variability in fishing time. A second cluster was the most numerous including 10 fishermen that have intermediate CVs and values in their proximity of fishing to the full moon. They also have high variability in effort, but lower than the previous cluster. Finally, the third cluster included nine fishermen with the highest CVs in yield and the highest frequency of fishing spent in proximity or during the full moon. They also had the lowest variability in fishing time.

GENERAL DISCUSSION AND FISHING PROFILES

The discussion is organized around three findings. First, in interviews and surveys fishermen recognized the importance of the traditional moon

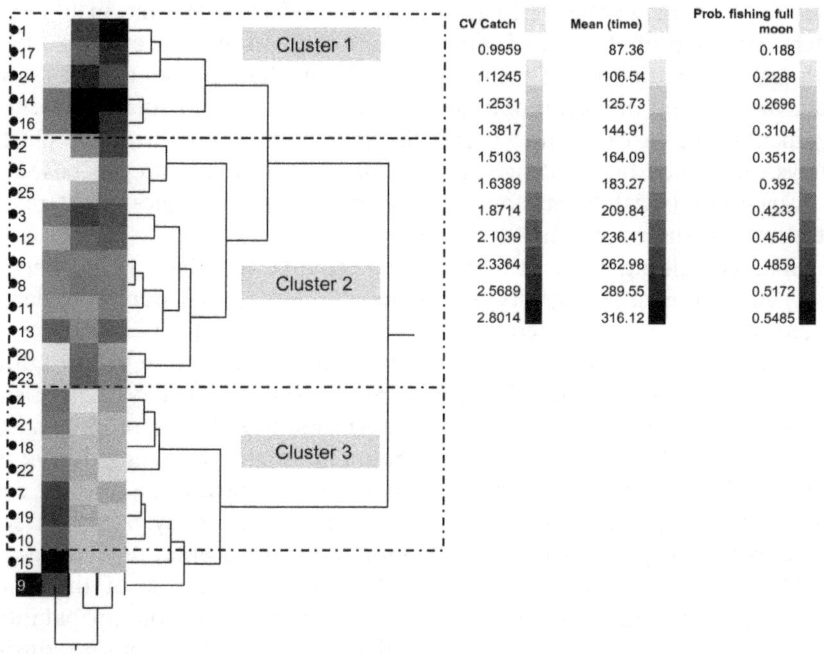

Clustering History

Number of Clusters	Distance	Leader	Joiner
24	0.160089617	6	8
23	0.326836616	7	19
22	0.334712184	14	16
21	0.349713299	6	11
20	0.391267945	4	21
19	0.413063428	20	23
18	0.422574111	7	10
17	0.496757342	5	25
16	0.527999979	1	17
15	0.616592987	3	12
14	0.637122135	4	18
13	0.737257023	1	24
12	0.743709660	4	22
11	0.821673999	6	13
10	0.886840209	2	5
9	1.015434449	7	15
8	1.190628895	3	6
7	1.327431992	7	9
6	1.457300015	3	20
5	1.774739408	1	14
4	2.161588857	4	7
3	2.284820259	2	3
2	3.266106325	1	2
1	6.100529793	1	4

Fig. 6. Dendrogram of Clusters of Fishermen Strategies. *Notes*: The dendrogram represents the ordering of subjects by CV in catches, mean of time spent fishing, and probability of fishing in full moon. The tree was cut in three clusters according to distances. Further detail is provided in the clustering history table with measured distances and order of inclusion.

calendar to allocate their effort. Second, the effect of the moon phase in catch rates or probability of fishing did not occur in the expected direction, with lower probabilities of fishing in the intermediate phase, fishing taking place also during the full moon, and differences between subjects playing a significant role in explaining variability in returns. And third, the cluster analysis indicated the existence of three different fishing strategies based on the variability of catch rates and time spent fishing. Strategies might reflect the effects of environmental uncertainty.

First, coinciding with ethnographic accounts (Cordell, 1974; Poepoe et al., 2007), the large majority of Endenese fishermen that were interviewed or participated in surveys considered the lunar cycle and its influences on tides and fishing returns as one of the main factors to regulate effort. According to the calendar, fishermen rest during the 10 days comprising the plenilune, and go fishing in the intermediate and new moon phases. They also time their trips to last 6–7 hours per night, departing during the ebb period and returning on the rising tide. In their narratives, nothing contradicts the idea that effort allocation decisions are mostly based on ecological beliefs originating in the perception of environmental regularities.

However, findings of regression analyses testing the role of the moon cycle in determining observed effort showed, instead, that the probabilities of going fishing were the lowest during the half-moon phase and not during the full moon. This is also observed in the frequency analysis, where fishing activity sometimes happens during the plenilune. Furthermore, the multilevel model studying the association between catch rates and the lunar cycle as stated in prediction 2 discovered that individual traits played a significant role in accounting for variability in the observations. As shown in the analyses, differences at the personal or individual level, like skills, traits, or preferences, might be a better explanatory candidate for effort allocation than the lunar cycle or the type of métier. Findings could be explained under the "skipper effect" (Acheson, 1981; King, 2011; Ruttan & Tyedmers, 2007; Thorlindsson, 1988). Yet, additional evidence on skill levels needs to be considered before that statement can be made.

The intercorrelation of measurements points to divergences in fishing preferences and time allocation that should not be there if all individuals were following the traditional moon calendar. Reasons why fishermen might choose not to rely on the lunar cycle might originate in the nature of fishing as an opportunistic activity. Because of the transient nature of sailing conditions, some anecdotal and ethnographic accounts mention that in small-scale fisheries decisions to go fishing are made at the moment of departure, and may not necessary include any long-term planning (e.g., Astuti, 1995). While throughout interviews Endenese fishermen singled out safety in navigation as the primary reason for not going fishing, it is also true that cluster analyses organized fishermen in three major groups. This clustering suggests that divergences in behavior are somewhat systematic, and even if the byproduct of cumulative opportunistic decisions, they might capture a reorganization of effort into a new configuration.

The reorganization of effort can be the symptom of higher uncertainty in environmental conditions. In fact, fishermen were very concerned about repeated changes in the ocean, weather, and climate conditions (Ramenzoni, 2013). Variations in oceanographic and atmospheric patterns, such as an increase in the temporality, frequency, and intensity of extreme events, have prevented fishermen from operating at expected times and created the incentive to develop new attitudes and practices toward fishing. For example, as expected by prediction 5, by fishing during the plenilune, fishermen show a high tolerance for variability in daily returns and the time they spend fishing. The dissonance between fishermen's narratives and actual effort can be explained by the unprecedented velocity of environmental changes, the scale in which such changes occur, and the dynamic nature of adaptation processes (Dessai et al., 2009). In relation to the scale and velocity of change, whereas fishermen are aware of irregularities that are forcing them to rearrange harvesting practices during seasonal cycles, they might not have the necessary evidence to detect a temporal pattern at the monthly level (Van Densen, 2001). As a consequence, the lunar calendar might not have been updated to reflect concurrent adjustments in effort.

Drawing from these points, I propose three profiles to classify effort strategies:

1. Conservative Time Allocation (TA) fishermen that minimize the number of days they spend fishing in intermediate phases and/or reduce the number of hours spent fishing during the full moon. These are fishermen that will show decreasing effort as days approach the full moon and increasing effort as dates move further from the full moon. They will strive to minimize the CV in returns per day, and they will have mid- to high CV in fishing times.
2. Opportunistic TA: fishermen that will fish on the full moon days more often to benefit from lower competition and higher prices. They might tolerate high variance in daily returns and they will have the lowest variability in fishing time.
3. Generalistic/flexible TA: fishermen that prioritize the allocation of time to fishing, regardless of the impact of the lunar cycle. These are

fishermen that will fish with the same probability on every single day of the month. They might tolerate high variance of returns per day, but, on average, have lower CV in catches than the opportunistic strategy. They also have mid- to high variability in fishing times.

Two considerations need to be made in relation to these profiles. First, profiles include fishermen disregarding of their métier. As analyses showed there is no significant difference in the CV in catches when comparing among métiers. Second, the fact that harvesting also occurs in the full moon for two of these profiles indicates that fishermen are willing to risk lower catches while maximizing the time spent fishing. Because returns are supposed to be lower (Cordell, 1974; Ramenzoni, 2013), fishing during the full moon increases the likelihood of coming home empty handed due to gear inefficiency. However, fishing during the full moon, if successful, will report greater monetary returns as competition is lower and prices of fish are higher than in other phases. Thus, even if catches are predicted to be smaller, fishermen might choose to go out on those nights to bring gains that are higher than normal days. Over the long run, this strategy might reduce variability in profits.

CONCLUSION

Through a mixed methods approach, the paper (1) investigated the use of the Endenese traditional lunar calendar; (2) determined whether lunar phases have a significant effect on fishing activities and catch rates; (3) identified profiles in temporal effort allocation that can be valuable for policy makers; and (4) discussed the role of environmental and economic uncertainties in shaping fishing behavior.

Considered as a whole, results contradict the traditional perception that Endenese fishermen's effort is strongly constrained by the full moon effect as in other fisheries (i.e., Cordell, 1974, Pet-Soede et al., 1999, 2001; van Densen, 2001; van Oostenbrugge et al., 2001). Although conversations and interviews among Endenese fishermen generally presented the lunar calendar as the main factor to determine effort, the analysis of behaviors among the 25 participants for over 9 months adds complexity to oral narratives and shows that time allocation strategies might differ from what is deemed ideal. This is a preliminary conclusion that requires further exploration but it suggests environmental uncertainty as a concrete driver of behavioral change. It is also a conclusion shared among some traditional fisheries in Australia, Samoa, and the Solomon Islands (Green et al., 2010; Hamilton & Walter, 1999; Lefale, 2010). In the case of Ende, causes of uncertainty can be traced back to environmental fluctuations as well as the interactions of ecological change with the effects of intensive exploitation.

Finally, this paper argues that an analysis that does not take into account the different kinds of variability in the dataset cannot provide good explanations of behavior (Álvarez & Schmidt, 2006). The aggregation of individual observations into a generic subject can make the elicitation of fishing profiles a difficult task by obscuring the detection of behavioral patterns (Salas et al., 2004). In this case, it could mean that the analyses of the factors affecting catch rates and effort allocation might be unable to capture the flexibility shown by responses and fully characterize the dynamic nature of adaptation processes (Dessai et al., 2009). In the past, taxonomies of resource users have been prone to misinterpretations by approaches that are strictly bioeconomic in their understanding of fishermen's decision-making and behavioral change. These perspectives have resulted in further policy failure when introduced in data-poor fishery management settings (Béné, 2003; Béné, Hersoug, & Allison, 2010; Coulthard, 2011). With constraints in technical and informational resources, regional governments have relied on rapid appraisal methods that, while providing snapshots of the situation, cannot capture its complexity (Béné & Tewfik, 2001). Although anthropological and sociological studies of fishing communities in transition could help solve some of these problems, the complex narrative of many ethnographic studies had encountered obstacles in its transition to applied management solutions (Hamilton & Walter, 1999). In all, there is a discord between highly specific qualitative and quantitative studies of fisheries and their instrumentation in policy. A multilevel approach to variability in fishermen's behaviors and motivations allows for the generation of strategies that can be used to model behavior. It constitutes a good option to systematize and address the heterogeneity that characterizes fisheries, making the adaptation of results to governance guidelines easier.

The exploration of the different kinds of variability and how individuals respond to change is crucial to designing sound policies. In Lopes and Begossi's words: "[d]esigning management initiatives without considering fishermen's individuality and personal goals that determine their behaviour means ignoring the most complex aspect of a chain that determines the degree of exploitation of a resource" (2011, p. 401). As fishermen adapt to new uncertain conditions, it is time economic anthropology and human behavioral ecology perspectives take on the task of studying individuals in their terms and not in averages.

LIMITATIONS AND FUTURE DIRECTIONS OF RESEARCH

Limitations

Significant limitations associated with this study include sample size and the representativeness of findings to the larger population. Another limitation is associated with the impact of development practices and socioeconomic changes in fishing strategies and how they might interrelate with environmental change to increase perceived uncertainty and, in turn, alter effort allocation (van Oostenbrugge, Van Densen, & Machiels, 2004). In relation to the first limitation, the proposed profiles are suggested as a heuristic that should be tested in future studies. In relation to the second limitation, the lack of industrial development and support from the fishing commission to intensify fishing operations implies that there are virtually no external incentives to maximize production and no alternative livelihoods to fishing. Socioeconomic pressures, through the loss of subsidies and high inflation, have had a dire impact on households. Migration has been the preferred solutions for younger generations, with women seeking employment in the Middle East and Malaysia. Future analyses will consider socioeconomic uncertainty and how it shapes effort allocation practices at the household level.

Future Directions

Observations and findings from this study seem to indicate that the most unstable days in terms of variability of returns occur on the calendar days (±4/5 days) before and after the full moon and also comprise part of the half-moon phase. In light of these results, there is a need for studies that can explore in more detail the application of the logistic function to capture the moon cycle. Results of binomial logistic models may vary depending on how the three classes representing the lunar cycle are constructed and which days are included in the intermediate and full moon phase.

In addition, changes in CVs in catches suggest a potential effect of the moon phase on the fishery. Further research should clarify if this difference is equal for all the species harvested.

Finally, more research is needed to assess the effect of catchability, seasonality, species types (aggregating and feeding behavior), and bathymetry in the variability of returns and how they associate with the moon cycle.

REFERENCES

Aburto, J., Thiel, M., & Stotz, W. (2009). Allocation of effort in artisanal fisheries: The importance of migration and temporary fishing camps. *Ocean & Coastal Management*, *52*(12), 646–654.

Acciaioli, G. (2004). From economic actor to moral agent: Knowledge, fate and hierarchy among the Bugis of Sulawesi. *Indonesia*, *78*, 147–179.

Acheson, J. M. (1981). Anthropology of fishing. *Annual Review of Anthropology*, *10*, 275–316.

Álvarez, A., & Schmidt, P. (2006). Is skill more important than luck in explaining fish catches? *Journal of Productivity Analysis*, *26*(1), 15–25.

Ammarell, G. (2002a). Bugis migration and modes of adaptation to local situations. *Ethnology*, *41*(1), 51–68.

Ammarell, G. (2002b). Knowing when to set sail: Practical knowledge and simple heuristics in Bugis navigational strategies. *Bijdragen Tot De Taal-, Land-En Volkenkunde*, *158*(2), 191–223.

Annamalai, H., & Liu, P. (2005). Response of the Asian summer monsoon to changes in El Niño properties. *Quarterly Journal of the Royal Meteorological Society*, *131*(607), 805–831.

Astuti, R. (1995). *People of the sea: Identity and descent among the Vezo of Madagascar*. Cambridge: Cambridge University Press.

Aswani, S. (1998). Patterns of marine harvest effort in southwestern New Georgia, Solomon Islands: Resource management or optimal foraging? *Ocean & Coastal Management*, 40(2), 207–235.

Badan Pusat Statistik Kabupaten Ende. (1984–2012). *Ende Dalam Angka*. Ende, Flores: Badan Pusat Statistik Kabupaten Ende.

Begossi, A. (1992). The use of optimal foraging theory in the understanding of fishing strategies: A case from Sepetiba Bay (Rio de Janeiro State, Brazil). *Human Ecology*, *20*(4), 463–475.

Bell, J. D., Ganachaud, A., Gehrke, P. C., Griffiths, S. P., Hobday, A. J., Hoegh-Guldberg, O., … Waycott, M. E. (2013). Mixed responses of tropical pacific fisheries and aquaculture to climate change. *Nature Climate Change*, *3*(6), 591–599.

Béné, C. (2003). When fishery rhymes with poverty: A first step beyond the old paradigm on poverty in small-scale fisheries. *World Development*, *31*(6), 949–975.

Béné, C., Hersoug, B., & Allison, E. H. (2010). Not by rent alone: Analysing the pro-poor functions of small-scale fisheries in developing countries. *Development Policy Review*, *28*(3), 325–358.

Béné, C., & Tewfik, A. (2001). Fishing effort allocation and fishermen's decision making process in a multi-species small-scale fishery: Analysis of the conch and lobster fishery in Turks and Caicos Islands. *Human Ecology*, *29*(2), 157–186.

Bjarnason, T., & Thorlindsson, T. (1993). In defense of a folk model: The "skipper effect" in the Icelandic cod fishery. *American Anthropologist*, *95*(2), 371–394.

Bolger, N., Davis, A., & Rafaeli, E. (2003). Diary methods: Capturing life as it is lived. *Annual Review of Psychology*, *54*(1), 579–616.

Bolger, N., & Laurenceau, J. P. (2013). *Intensive longitudinal methods: An introduction to diary and experience sampling research*. New York, NY: Guilford Press.

Bos, A. R., & Gumanao, G. S. (2012). The lunar cycle determines availability of coral-reef fishes at fish markets. *Journal of Fish Biology*, *81*(6), 2074–2079.

Branch, T. A., Hilborn, R., Haynie, A. C., Fay, G., Flynn, L., Griffiths, J., & Ward, E. J. (2006). Fleet dynamics and fishermen behavior: Lessons for fisheries managers. *Canadian Journal of Fisheries and Aquatic Sciences*, *63*(7), 1647–1668.

Chakravarty-Kaul, M. (1998). Transhumance and customary pastoral rights in Himachal Pradesh: Claiming the high pastures for Gaddis. *Mountain Research and Development*, *18*, 5–17.

Cheung, W. W. L., Lam, V. W. Y., Sarmiento, J. L., Kearney, K., Watson, R., & Pauly, D. L. (2009). Projecting global marine biodiversity impacts under climate change scenarios. *Fish and Fisheries*, *10*(3), 235–251.

Cnaan, A., Laird, N. M., & Slasor, P. (1997). Tutorial in biostatistics: Using the general linear mixed model to analyse unbalanced repeated measures and longitudinal data. *Statistics in Medicine*, *16*(20), 2349–2380.

Cordell, J. (1974). The lunar-tide fishing cycle in Northeast Brazil. *Ethnology*, *13*, 379–392.

Coulthard, S. (2011). Poverty, sustainability and human wellbeing: A social wellbeing approach to the global fisheries crisis. *Global Environmental Change*, *21*(2), 453.

Daw, T., Adger, W. N., Brown, K., & Badjeck, M. (2009). Climate change and capture fisheries: Potential impacts, adaptation and mitigation. *Climate change implications for fisheries and aquaculture: Overview of current scientific knowledge*. FAO Fisheries and Aquaculture Technical Paper No. 530, 107–150.

Dayem, K. E., Noone, D. C., & Molnar, P. (2007). Tropical western pacific warm pool and maritime continent precipitation rates and their contrasting relationships with the walker circulation. *Journal of Geophysical Research: Atmospheres (1984–2012)*, *112*(D6). doi:10.1029/2006JD007870

Defeo, O., & Castilla, J. C. (2005). More than one bag for the world fishery crisis and keys for co-management successes in selected artisanal Latin American shell fisheries. *Reviews in Fish Biology and Fisheries*, *15*(3), 265–283.

Demanèche, S., Merrien, C., Vigneau, J., Guyader, O., Berthou, P., Lespagnol, P., … Daurès, F. (2008). A new approach to estimate catches and fishing effort of small scale fisheries by sampling fishing trips on-site. Paper presented at Proceedings of the ICES Annual Science conference.

Dessai, S., Hulme, M., Lempert, R., & Pielke, R., Jr. (2009). Climate prediction: A limit to adaptation. *Adapting to Climate Change: Thresholds, Values, Governance* (pp. 64–78). Cambridge: Cambridge University Press.

Di Natale, A., & Mangano, A. (1995). Moon phases influence on CPUE: A first analysis of swordfish driftnet catch data from the Italian fleet between 1990 and 1991. *Collective Volume of Scientific Papers ICCAT*, *44*(1), 264–267.

Dietrich, S. (1983). Flores in the nineteenth century: Aspects of Dutch colonialism on a non-profitable island. *Indonesia Circle*, *11*(31), 39–58.

Durrenberger, E., & Pálsson, G. (1986). Finding fish: The tactics of Icelandic skippers. *American Ethnologist*, *13*(2), 213–229.

Firth, R. (1965 [2006]). *Malay fishermen*. London, Routledge.

García-Quijano, C. G. (2006). *Resisting extinction: The value of local ecological knowledge for small-scale fishers in Southeastern Puerto Rico*. Dissertation. The University of Georgia.

Godoy, R. A., Reyes-García, V., Vadez, V., Magvanjav, O., Leonard, W. R., McDade, T., & ... TAPS Bolivia Study Team (2009). Does culture affect the present? The effect of future weather on the current collection of planted crops and wildlife in a native Amazonian society of Bolivia. *Human Ecology*, *37*(5), 613–628.

Green, D., Billy, J., & Tapim, A. (2010). Indigenous Australians' knowledge of weather and climate. *Climatic Change*, *100*(2), 337–354.

Guyader, O., Berthou, P., Koutsikopoulos, C., Alban, F., Demaneche, S., Gaspar, M. B., ... Maynou, F. (2013). Small scale fisheries in Europe: A comparative analysis based on a selection of case studies. *Fisheries Research*, *140*, 1–13.

Halim, A. (2002). Adoption of cyanide fishing practice in Indonesia. *Ocean and Coastal Management*, *45*, 313–323.

Hamilton, R., & Walter, R. (1999). Indigenous ecological knowledge and its role in fisheries research design: A case study from Roviana Lagoon, Western Province, Solomon Islands. *SPC Traditional Marine Resource Management and Knowledge Information Bulletin*, *11*, 13–25.

Hamon, K. (2009). Typology and dynamics of a spatially and temporally heterogeneous fishery: The Tasmanian rock lobster fishery. Paper presented at 18th World Imacs Congress and Modsim09 International Congress on Modelling and Simulation-Interfacing Modelling and Simulation with Mathematical and Computational Sciences.

Hayes, A. F. (2006). A primer on multilevel modeling. *Human Communication Research*, *32*(4), 385–410.

Hendon, H. H. (2003). Indonesian rainfall variability: Impacts of ENSO and local Air–Sea interaction. *Journal of Climate*, *16*, 1775–1790.

King, T. J. (2011). The "skipper effect": Riddles of luck and rhetorics of individualism. *Human Organization*, *70*(4), 387–396.

Koster, J. M., Grote, M. N., & Winterhalder, B. (2013). Effects on household labor of temporary out-migration by male household heads in Nicaragua and Peru: An analysis of spot-check time allocation data using mixed-effects models. *Human Ecology*, *41*(2), 221–237.

Lefale, P. F. (2010). Ua 'afa le aso stormy weather today: Traditional ecological knowledge of weather and climate. The Samoa experience. *Climatic Change*, *100*(2), 317–335.

Lehodey, P., Bertignac, M., Hampton, J., Lewis, A., & Picaut, J. (1997). El Niño southern oscillation and tuna in the Western Pacific. *Nature*, *389*(6652), 715–718.

Lehodey, P., Chai, F., & Hampton, J. (2003). Modelling climate-related variability of tuna populations from a coupled ocean–biogeochemical-populations dynamics model. *Fisheries Oceanography*, *12*(4–5), 483–494.

Lopes, P. F. M., & Begossi, A. (2011). Decision-making processes by small-scale fishermen on the southeast coast of Brazil. *Fisheries Management and Ecology*, *18*(5), 400–410.

Lopes, P. F. M., Clauzet, M., Hanazaki, N., Ramires, M., Silvano, R. A. M., & Begossi, A. (2011). Foraging behaviour of Brazilian riverine and coastal fishers: How much is explained by the optimal foraging theory? *Conservation & Society*, *9*(3), 236–246. doi:10.4103/0972-4923.86994

Lopeztegui-Castillo, A., Pinar, N. C., & Baisre Álvarez, J. A. (2013). El ciclo lunar y las capturas de langosta Panulirus Argus (*Decapoda: Palinuridae*) en la zona de Caibarien, Cuba. *Boletin del Centro de Investigaciones Biologicas*, *45*(2), 131–142.

Lowe, C. (2006). *Wild profusion*. Princeton, NJ: Princeton University Press.

Mahévas, S., Bellanger, L., & Trenkel, V. M. (2008). Cluster analysis of linear model coefficients under contiguity constraints for identifying spatial and temporal fishing effort patterns. *Fisheries Research*, *93*(1), 29–38.

Malinowski, B. (1918). Fishing in the Trobriand Islands. *Man*, *18*, 87–92.

Malinowski, B. (1921). The primitive economics of the Trobriand islanders. *The Economic Journal*, *13*, 1–16.

Malinowski, B. (1978). *Argonauts of the Western Pacific: An account of native enterprise and adventure in the archipelagoes of Melanesian New Guinea*. London, Routledge.

Marchal, P., Poos, J. J., & Quirijns, F. (2007). Linkage between fishers' foraging, market and fish stocks density: Examples from some North Sea fisheries. *Fisheries Research*, *83*(1), 33–43.

McElreath, R., & Koster, J. (2014). Using multilevel models to estimate variation in foraging returns. *Human Nature*, *25*(1), 100–120.

Millar, R. B., McKenzie, J. E., Bell, J. D., & Tierney, L. D. (1997). Evaluation of an indigenous fishing calendar using recreational catch rates of snapper *Pagrus auratus* in the north island of New Zealand. *Marine Ecology Progress Series*, *151*(1–3), 219–224.

Millar, R. B., & Willis, T. J. (1999). Applications: Estimating the relative density of snapper in and around a marine reserve using a log-linear mixed-effects model. *Australian & New Zealand Journal of Statistics*, *41*(4), 383–394.

Morrill, W. T. (1967). Ethnoicthyology of the Cha-Cha. *Ethnology*, *6*, 405–416.

Mous, P. J., Pet, J. S., Arifin, Z., Djohani, R., Erdmann, M. V., Halim, A., ... Wiadnya, G. (2005). Policy needs to improve marine capture fisheries management and to define a role for marine protected areas in Indonesia. *Fisheries Management and Ecology*, *12*(4), 259–268.

Nakagawa, S. (1984). Endenese. In R. V. Weekes (Ed.), *Muslim peoples: A world ethnographic survey* (pp. 251–253). Westport, CN: Greenwood Press.

Needham, R. (1968). Endeh: Terminology, alliance, and analysis. *Bijdragen Tot De Taal-, Land-En Volkenkunde*, *124*, 305–335.

Needham, R. (1980). Principles and variations in the structure of the Sumbanese society. In J. J. Fox & M. Adams (Eds.), *The flow of life: Essays on Eastern Indonesia* (pp. 21–47). Boston, MA: Harvard Press.

Orlove, B. S., Chiang, J. C. H., & Cane, M. A. (2002). Ethnoclimatology in the Andes. *American Scientist*, *90*, 428–435.

Ortega-García, S., Ponce-Diaz, G., O'Hara, R., & Merilä, J. (2008). The relative importance of lunar phase and environmental conditions on striped marlin (*Tetrapturus audax*) catches in sport fishing. *Fisheries Research*, *93*(1), 190–194.

Ott, L., & Longnecker, M. T. (2010). *An introduction to statistical methods and data analysis*. Belmont, CA: Brooks/Cole, Cengage Learning.

Paolisso, M. (2002). Blue crabs and controversy on the Chesapeake Bay: A cultural model for understanding watermen's reasoning about Blue Crab management. *Human Organization*, *61*(3), 226–239.

Pasaribu, S. M. (2007). Revitalizing institutions to enhance climate forecast application in East Nusa Tenggara province, Indonesia. *Anal Kebijak Pertan*, *5*(3), 239–253.

Pelletier, D., & Ferraris, J. (2000). A multivariate approach for defining fishing tactics from commercial catch and effort data. *Canadian Journal of Fisheries and Aquatic Sciences*, *57*(1), 51–65.

Pet, J. S., Van Densen, W. L. T., Machiels, M. A. M., Setyohadi, D., & Tumuljadi, A. (1997). Catch, effort and sampling strategies in the highly variable sardine fisheries around East Java, Indonesia. *Fisheries Research*, *31*(1), 121–137.

Pet-Soede, C., Van Densen, W. L. T., Hiddink, J. G., Kuyl, S., & Machiels, M. A. M. (2001). Can fishermen allocate their fishing effort in space and time on the basis of their catch rates? An example from Spermonde archipelago, SW Sulawesi, Indonesia. *Fisheries Management & Ecology*, *8*(1), 15–36.

Pet-Soede, L., Cesar, H., & Pet, J. (2000). Blasting away: The economics of blast fishing on Indonesian coral reefs. In H. Cesar (Ed.), *Collected essays on the economics of coral reefs* (pp. 77–84). Kalmar: CORDIO.

Poepoe, K., Bartram, P., & Friedlander, A. (2007). The use of traditional Hawaiian knowledge in the contemporary management of marine resources. In N. Haggan, B. Neis, & I. G. Baird (Eds.), *Fishers' knowledge in fisheries science and management* (pp. 117–141). Paris: UNESCO.

Poisson, F., Gaertner, J.-C., Taquet, M., Durbec, J. C., & Bigelow, K. (2010). Effects of lunar cycle and fishing operations on longline-caught pelagic fish: Fishing performance, capture time, and survival of fish. *Fishery Bulletin*, *108*, 268–281.

Potemra, J. T., Hautala, S. L., & Sprintall, J. (2003). Vertical structure of Indonesian through flow in a large-scale model. *Deep Sea Research Part II: Topical Studies in Oceanography*, *50*(12), 2143–2161.

Ramenzoni, V. C. (2013). Endenese fisheries: Exploratory findings on environmental perceptions, fish effort, and overfishing in eastern Indonesia. *Ethnobiology Letters*, *4*, 39–51.

Roessig, J. M., Woodley, C. M., Cech, J. J. Jr., & Hansen, L. J. (2004). Effects of global climate change on marine and estuarine fishes and fisheries. *Reviews in Fish Biology and Fisheries*, *14*(2), 251–275.

Ruttan, L. M., & Tyedmers, P. H. (2007). Skippers, spotters and seiners: Analysis of the "skipper effect" in US menhaden (*Brevoortia*) purse-seine fisheries. *Fisheries Research*, *83*(1), 73–80.

Salas, S., & Charles, A. (2007). Are small-scale fishers profit maximizers?: Exploring fishing performance of small-scale fishers and factors determining catch rates. *Proceedings of the 60th Gulf and Caribbean Fisheries Institute*.

Salas, S., & Gaertner, D. (2004). The behavioural dynamics of fishers: Management implications. *Fish and Fisheries*, *5*(2), 153–167.

Salas, S., Sumaila, U. R., & Pitcher, T. (2004). Short-term decisions of small-scale fishers selecting alternative target species: A choice model. *Canadian Journal of Fisheries & Aquatic Sciences*, *61*(3), 374–383.

Sethi, G., Costello, C., Fisher, A., Hanemann, M., & Karp, L. (2005). Fishery management under multiple uncertainty. *Journal of Environmental Economics and Management*, *50*(2), 300–318.

Silvano, R. A. M., & Begossi, A. (2001). Seasonal dynamics of fishery at the Piracicaba River (Brazil). *Fisheries Research*, *51*(1), 69–86.

Snijders, T. A. B. (2011). *Multilevel analysis*. New York, NY, Springer.

Snijders, T. A. B., & Bosker, R. J. (1999). *Multilevel analysis: An introduction to basic and advanced multilevel modeling*. London: Sage.

Sosis, R. (2001). Sharing, consumption, and patch choice on Ifaluk atoll. *Human Nature*, *12*(3), 221–245.

Sosis, R. (2002). Patch choice decisions among Ifaluk fishers. *American Anthropologist*, *104*(2), 583–598.

Srisurichan, S., Caputi, N., & Cross, J. (2005). Impact of lunar cycle and swell on the daily catch rate of western rock lobster (*Panulirus cygnus*) using time series modelling. *New Zealand Journal of Marine and Freshwater Research*, *39*, 749–764.

Stacey, N., Nurhakim, S., Nugroho, D., Soselisa, H., Resosudarmo, B., Kalis, O., … Karam, J. (2011). *Socio-economic profile of the Arafura and Timor seas*. Report Prepared for the Arafura Timor Seas Ecosystem Action (ATSEA) Program. Jakarta: ATSEA.

Stelzenmüller, V., Maynou, F., Bernard, G., Cadiou, G., Camilleri, M., Crec'hriou, R., … Higgins, R. (2008). Spatial assessment of fishing effort around European marine reserves: Implications for successful fisheries management. *Marine Pollution Bulletin*, *56*(12), 2018–2026.

Stephens, D. W. (2008). Decision ecology: Foraging and the ecology of animal decision making. *Cognitive, Affective, & Behavioral Neuroscience*, *8*(4), 475–484.

Stevenson, B. C., & Millar, R. B. (2013). Promising the moon? Evaluation of indigenous and lunar fishing calendars using semiparametric generalized mixed models of recreational catch data. *Environmental and Ecological Statistics*, *20*(4), 591–608.

Stocker, T. F., Qin, D., Plattner, G. K., Tignor, M., Allen, S. K., Boschung, J., & IPCC (2013). Climate change 2013: The physical science basis. Contribution of working group I to the fifth assessment report of the intergovernmental panel on climate change.

Teh, L. C. L., Teh, L. S. L., & Meitner, M. J. (2012). Preferred resource spaces and fisher flexibility: Implications for spatial management of small-scale fisheries. *Human Ecology*, *40*(2), 213–226. doi:10.1007/s10745-012-9464-9

Thorlindsson, T. (1988). The skipper effect in the Icelandic herring fishery. *Human Organization*, *47*(3), 199–212.

TROFCCA. (2006, May 29–30). Climate change projections for Indonesia. Background document for the Southeast Asia kick-off meeting of the project Tropical Forests and Climate Change Adaptation ("TroFCCA"). Bogor.

Tucker, B. (2007). Perception of interannual covariation and strategies for risk reduction among Mikea of Madagascar. *Human Nature*, *18*, 162–180.

Tzanatos, E., Somarakis, S., Tserpes, G., & Koutsikopoulos, C. (2006). Identifying and classifying small-scale fisheries métiers in the Mediterranean: A case study in the Patraikos Gulf, Greece. *Fisheries Research*, *81*, 158–168.

van Densen, W. L. T. (2001). *On the perception of time trends in resource outcome: Its importance in fisheries co-management, agriculture and whaling*. Enschede, The Netherlands: University of Twente.

van der Kaars, S., Bassinot, F., De Deckker, P., & Guichard, F. (2010). Changes in monsoon and ocean circulation and the vegetation cover of southwest Sumatra through the last 83,000 years: The record from marine core BAR94-42. *Palaeogeography, Palaeoclimatology, Palaeoecology*, *296*, 52–78. doi:10.1016/j.palaeo.2010.06.015

van Oostenbrugge, J. A. E. (2001). *Uncertainty in daily catch rate in the light fisheries around Ambon and the lease islands: Characterisation, causes and consequences[Sl: sn]*. Dissertation.

van Oostenbrugge, J. A. E., Bakker, E. J., van Densen, W. L. T., Machiels, M. A. M., & van Zwieten, P. A. M. (2002). Characterizing catch variability in a multispecies fishery: Implications for fishery management. *Canadian Journal of Fisheries and Aquatic Sciences*, *59*(6), 1032–1043.

van Oostenbrugge, J. A. E., Densen, W. L. T., & Machiels, M. A. M. (2001). Risk aversion in allocating fishing effort in a highly uncertain coastal fishery for pelagic fish, Moluccas, Indonesia. *Canadian Journal of Fisheries and Aquatic Sciences*, *58*(8), 1683–1691.

van Oostenbrugge, J. A. E., Van Densen, W. L. T., & Machiels, M. A. M. (2004). How the uncertain outcomes associated with aquatic and land resource use affect livelihood strategies in coastal communities in the central Moluccas, Indonesia. *Agricultural Systems*, *82*(1), 57–91.

van Zwieten, P. A. M., Roest, F. C., Machiels, M. A. M., & Van Densen, W. L. T. (2002). Effects of inter-annual variability, seasonality and persistence on the perception of long-term trends in catch rates of the industrial pelagic purse-seine fishery of northern Lake Tanganyika (Burundi). *Fisheries Research*, *54*(3), 329–348.

Vázquez-Rowe, I., & Tyedmers, P. (2013). Identifying the importance of the "skipper effect" within sources of measured inefficiency in fisheries through data envelopment analysis (DEA). *Marine Policy*, *38*, 387–396.

Westneat, D., & Fox, C. W. (2010). *Evolutionary behavioral ecology*. Oxford: Oxford University Press.

Winterhalder, B., & Smith, E. A. (2000). Analyzing adaptive strategies: Human behavioral ecology at twenty-five. *Evolutionary Anthropology Issues News and Reviews*, *9*(2), 51–72.

Wiyono, E. S., Yamada, S., Tanaka, E., Arimoto, T., & Kitakado, T. (2006). Dynamics of fishing gear allocation by fishers in small-scale coastal fisheries of Pelabuhan Ratu Bay, Indonesia. *Fisheries Management & Ecology*, *13*(6), 185–195.

Zeller, D., Booth, S., Davis, G., & Pauly, D. (2007). Re-estimation of small-scale fishery catches for US flag-associated island areas in the western pacific: The last 50 years. *Fishery Bulletin*, *105*(2), 266–277.

Zeller, D., Booth, S., & Pauly, D. (2006). Fisheries contributions to the gross domestic product: Underestimating small-scale fisheries in the Pacific. *Marine Resource Economics*, *21*(4), 355–374.

PART III
URBAN SITUATIONS

THE ORIGINS OF THE SUSTAINABILITY CONCEPT: RISK PERCEPTION AND RESOURCE MANAGEMENT IN EARLY URBAN CENTERS

Monica L. Smith

ABSTRACT

Purpose – *This paper examines the conditions under which ancient peoples might have developed a concept of "sustainability," and concludes that long-term resource management practices would not have been articulated prior to the development of the first cities starting c. 6,000 years ago.*

Methodology/approach – *Using biological concepts of population density and niche-construction theory, cities are identified as the first places where pressures on resources might have triggered concerns for sustainability. Nonetheless, urban centers also provided ample opportunities for individuals and households to continue the same ad hoc foraging strategies that had facilitated human survival in prior eras.*

Climate Change, Culture, and Economics: Anthropological Investigations
Research in Economic Anthropology, Volume 35, 215–238

ISSN: 0190-1281/doi:10.1108/S0190-128120150000035009

Social implications – *The implementation of a sustainability concept requires two things: individual and institutional motivations to mitigate collective risk over the long term, and accurate measurement devices that can discern subtle changes over time. Neither condition was applicable to the ancient world. Premodern cities provided the first expression of large population sizes in which there were niches of economic and social mutualism, yet individuals and households persisted in age-old approaches to provisioning by opportunistically using urban networks rather than focusing on a collective future.*

Originality/value – *Archaeological and historical analysis indicates that a focus on "sustainability" is not an innate human behavioral capacity but must be specifically articulated and taught.*

Keywords: Urbanism; risk; environment; institutions; infrastructure

INTRODUCTION

The concept of "sustainability" has garnered considerable contemporary interest as a rubric for the management of populations and extractive resource activities, often highlighted by a concern that "The world's present development path is not sustainable" (Kates et al., 2001, p. 1; see also Blaikie & Brookfield, 1987; Redman, 2014; White, 2010). Archaeologists have enthusiastically engaged in the literature of sustainability and human–environmental dynamics, suggesting that "knowledge generated by archaeology can play a unique and valuable role in developing the tools to make more informed decisions that will shape our future" (Barton, Ullah, Bergin, Mitasova, & Sarjoughian, 2012, p. 42; see also Costanza et al., 2012; Ellis et al., 2013, p. 7983; Kintigh et al., 2014, p. 12; Minnis, 2010; Redman, 1999, 2014, p. 5, M.E. Smith, 2010). The implication is that ancient peoples' approaches to the environment provide models for contemporary human action, whether through examples of human–environmental balance or through cautionary tales elicited from ecological disaster (summarized in Hames, 2007).

But when was "sustainability" first recognized and articulated? Did ancient humans also perceive that they overused or were in danger of overusing their environments with reference to past experiences and future expectations? Under what circumstances were environmental conditions and human–environmental interactions viewed as deleterious or in need of change? Environmental fluctuations in both slow and rapid forms were certainly perceptible, necessitating human response and accommodation. Rapid fluctuations can be caused by earthquakes, tsunamis, volcanoes, landslides, forest fires, hurricanes, oceanic storms, tornadoes, warfare, floods, and El Niño events, all of which entail changes of unpredictable duration, magnitude, and effect. Some of these large-scale changes have precursors on the scale of days or weeks, and others arrive without warning, resulting in an inability to plan for anything but a post hoc response. There also have been a number of slow changes over the course of the past 1.8 million years, including rises in sea level and the inception of the Holocene era in which climate became warmer and drier (di Lernia, 2006; Richerson, Boyd, & Bettinger, 2001).

Through periods of both rapid and slow change, an awareness of human–environmental dynamics would have been most obviously discerned when there was a disbalance between supply and demand. Population and carrying capacity are directly interrelated, such that areas of high population density are the likely locations of human–environmental challenges (Browman, 1987; Lambin et al., 2001). Although high population densities can be found in rural areas (e.g., a small mountain valley with densely distributed villages), cities were the first places where populations became sufficiently large to put consistent pressure on resources. The first cities were formed c. 6,000 years ago and faced the same opportunities and constraints as their modern-day counterparts. Because cities, in general, must acquire resources from outside of their boundaries (particularly, food and fuel produced in the surrounding hinterlands), city dwellers were the ones most likely to have first conceptualized ideas of "sustainability" as they assessed their past, present, and future access to needed resources for both daily needs and for the long term. Pressures were augmented by the fact that urban residents were for the most part not full-time farmers, and thus were reliant on attenuated networks of provisioning through markets, redistributive institutions, and/or direct reliance on kin in the countryside.

The objective potential for resource-acquisition stress in ancient cities need not have been perceived that way by their inhabitants, however, given that there is no simple equation between population density and perceptions of environmental stress. Blaikie and Brookfield (1987, p. 13) note that concepts such as degradation and resilience are social terms incumbent on observations of natural and cultural conditions, and circumstances have to first be identified and conceptualized as a "problem" before they can be addressed (see also Anscheutz, Wilshusen, & Scheick, 2001; Colten, 2005, p. 12; Mustafa, 2005). Even when phenomena such as pollution, crowding, waste management, or resource shortfall are recognized as "problems," they still may be viewed selectively because "One person's degradation is another's accumulation ..." (Blaikie & Brookfield, 1987, p. 14; see also Browman, 1987; Marcuse, 1998). And even if a phenomenon is defined as a "problem," there may not be any solution that can be identified or proposed by those who are affected (Douglas et al., 2008).

THE IDEA OF SUSTAINABILITY

Sustainability has been defined as the search for a balance between the present and future use of resources. One of the first articulations of this concept was by the forester Gifford Pinchot in 1910, who noted "our obligation ... to use what we need [such] that our descendants shall not be deprived of what they need" (cited in Owens, 2003, p. 6; see also Redman, 1999, p. 122, 2014). The concept of sustainability faces numerous challenges of implementation, however, even in the twenty-first century. Does sustainability mean that we should seek to manage our environment in order to restore prior conditions, to stabilize our current use, or to plan for anticipated needs at target thresholds of population and resources? If sustainability is to be measured against future targets, what levels of population, land-use, and food types are to be modeled and anticipated? Who is to be tasked with maintaining the trajectory to the target threshold, what incentives can be utilized to encourage or force compliance, and can thresholds be changed in light of new developments in technology that mitigate, reverse, or exacerbate prior adverse effects?

An additional challenge to the actualization of sustainability goals is the time frame for desired future outcomes. Sustainability today is assessed somewhere between one and four centuries in the future, which is an infinitesimally small slice of geologic time but well beyond the lifespan of an individual (cf. Weitzman, 1998). Although people can and do conceptualize long-term outcomes on the multigenerational scale, human beings appear to be better-equipped to assess short-term outcomes (on the scale of hours and days) and, to a certain extent, medium-term outcomes (on the scale of several years or an individual lifetime, measured by one's own growth and decline, as well as by the growth trajectory of offspring). The recognition of cognitive challenges to goals on a longer scale is found in various literatures dealing with the psychology of human behavior, ranging from daily acts of hygiene to the sequestering of funds for retirement and old age (e.g., Dubas & Jonsson, 2005; Rise, Åstrøm, & Sutton, 1998).

The observed propensity to focus on short-term and medium-term eventualities may be the result of the evolution of human cognitive limits: with short life spans until the modern period, there were few benefits to expending energy on unknown future outcomes. The time-scale of risks perceived by individuals can be linked to the notion of "future discounting" in economics, in which greater value is placed on the here-and-now rather than on the future (Weitzman, 1998). Instead of direct planning, which would have required knowledge of multivariate elements of environmental management and the capacity to proactively alter trajectories of use, our human ancestors instead relied on five other strategies to mitigate the potential of future loss. The first strategy was the propagation of large families to ensure the biological survival of at least some descendants who would also rely on each other in times of need through an extensive kin-network. The second strategy was the use of memory and information-sharing to stockpile knowledge that could be of use in case of future need; information is cheap, easily disseminated, and storable in ways that the physical storage of goods or food cannot be. A third strategy was "social storage" in which individuals share resources in anticipation of recalling those obligations in future (Halstead & O'Shea, 1989, p. 4). A fourth strategy was the development and use of physical technologies to resolve problems as they appeared (Boyd, Richerson, & Henrich, 2013). A final strategy was to develop new social traditions when crises occurred such as new forms of communal sharing or entitlements to scarce resources (di Lernia, 2006; Rautman, 2013; M. L. Smith, 2013). All of these strategies can be categorized as "real-time" or "just-in-time" strategies that yielded immediate results in the present as well as serving as insurance against future need.

THE ORIGINS OF ENVIRONMENTAL PERCEPTION

Prior to the Holocene, our ancestors were mobile foragers who migrated through landscapes engaging in cycles of resource acquisition in which adaptability and omnivory facilitated the successful use of many different environmental niches (Zeder, 2012). Resource differentiation was perceived both in spatial terms across landscapes and as in-place fluctuations over time, but conspecific competition for resources was relatively low. In a landscape that was distinctly underpopulated compared to today, individuals could personally identify and consistently engage with other known parties to identify, address, and forestall the known ranges of existing risk. Interactions with strangers may have taken place relatively infrequently, and in the context of several hundred people at most in which gatherings were primarily focused on ritual and mating opportunities (see Layton, O'Hara, & Bilsborough, 2012).

In the course of their movements, ancient peoples did recognize both the detrimental and generative effects of human agency on their surroundings and had the capacity to affect their local environments on a massive scale. One of the first widespread human technologies was fire, which was used to clear vegetation, drive game animals into traps, and improve soil fertility (Bliege Bird, Bird, Codding, Parker, & Jones, 2008; Roos, Sullivan, & McNamee, 2010). Hunting technologies such as improved projectile points enabled humans to become the apex predator in many novel environments and a likely factor in the extirpation of species in the Americas, Australia, and many islands (Loehle & Eschenbach, 2011; Prado, Arroyo-Cabrales, Johnson, Alberdi, & Polaco, 2012). Long before the development of full-time agriculture, foragers assisted plants by watering, tending them, or providing appropriate conditions for natural growth (B. Smith, 2001). Although the effects of such actions could be perceived by those other than the instigators, nearly all of these strategies were individual and household ones in which the focus of actions and consequences was the immediate present and short-term future.

Collective human effects on landscapes became more widespread starting c. 10–12,000 years ago in the Holocene (Ellis et al., 2013; Richerson et al., 2001), in which the adoption of plant and animal husbandry entailed significant social trade-offs including settled village life. Household mobility was constrained both through positive feedback (control over resources through accumulation) and negative feedback (inability to take over others' land and resources due to concepts of usufruct and ownership). The relationship between people and the groups in which they found themselves, always a component of human–environmental dynamics, became increasingly intertwined as population sizes increased. As households stayed in place to labor in fields and to safeguard harvested produce, they invested in more substantial dwellings, more extensive storage structures, and permanent cooking facilities that often involved heavy grinding equipment and immoveable features such as ovens. They also modified their landscapes and reacted to localized events of erosion and soil degradation through the construction of terraces and other remedial features (Borejsza, Rodríguez

López, Frederick, & Bateman, 2008). These individual and household actions were accompanied by the concomitant development of small-scale political hierarchies and more elaborate rituals of mutual aid and interaction including feasting and other forms of food redistribution (Bogaard et al., 2009; Hegmon, 1991; M. L. Smith, 2010).

Human actions have only recently been pervasive enough to result in measurable changes, as seen in ice-core studies that show an increase in methane gas produced through rice farming and animal husbandry starting 4,000 years ago (Fuller et al., 2011) and atmospheric pollutants that were the result of large-scale lead mining in the Carthaginian and Roman periods starting c. 600 BC (Rosman, Chisholm, Hong, Candelone, & Boutron, 1997). The broader global impacts of these practices, visible to us today through the use of measurement devices on longitudinal samples, were unlikely to have been perceived by ancient peoples except at a very localized level such as through the health effects on individuals working in those industries. Examples of ancient measuring devices for large-scale environmental variability are extremely rare, such as the so-called Nilometers of Egypt which consisted of both portable devices and marks carved into living rock, the measurements of which were recorded in royal annals starting with the First Dynasty c. 3050 BC (Bell, 1970) and coincident with initial expressions of regional political authority.

The Nilometer example illustrates the role of cultural awareness as an essential starting-point in perceptions of environmental variability. Political authorities needed to measure the Nile not as a matter of general curiosity but to predict agricultural yields for taxation and provisioning purposes. Elsewhere in the world, other types of "problems" became defined as populations increased, particularly in urban centers where the opportunities and constraints of shared landscape management were heightened. Even simple everyday acts such as the disposal of human waste and the discard of trash impinged on others' proximate use of space in ways that were not the case in village environments. As a result, cities became places where we can see a transition from simple cooperative behavior to more complex collective actions (cf. Carballo, Roscoe, & Feinman, 2014), manifested in the physical organization of urban spaces. For ancient urban dwellers, there were many arenas of interdependence between rural and urban realms related to gross inputs of food, water, and fuel, and many points of intersection among residents related to the net acquisition of raw ingredients and cooked meals, as well as spaces for social interactions, ritual activity, and economic pursuits. The feelings of competition for any of these elements introduced to urban inhabitants a level of stress and anxiety about procuring needed resources, such that a concept of "sustainability" can be envisioned as potentially applying to the planning for future resource acquisitions and maintenance to preserve investments in household and community welfare.

CITIES AND THEIR ENVIRONMENTS

For the first time in human history, cities represented places where large numbers of people willingly traded off direct control of their food supplies for a reliance on indirect provisioning through intermediaries. Rather than foraging directly for needed resources or storing a harvest's worth of grain within the household, individuals acquired small, regular distributions of food through a variety of mechanisms ranging from a reliance on rural family members, to wage labor compensated in kind or converted into comestibles through market mechanisms, to redistributions from authorities housed in temples or palaces. Although the development of often-indirect forms of resource acquisition required a "leap of faith" that should have been accompanied by an increased sense of risk and anxiety, cities in their diversity of provisioning systems may actually have represented reduced risk for individual inhabitants because they reinforced strategies of foraging to which our species had long been adapted.[1] In cities, individuals were able to simulate mobile hunting-and-gathering through the use of interdigitated economic, social, and institutional networks that provided "foraging" opportunities.

Cities represent accelerated patterns of production, consumption, distribution, and discard in which the increased scale of the economy provides employment for both skilled professionals and manual laborers (cf. Iyenda, 2005; Kemper, 2010; Sanders, Nee, & Sernau, 2002). Urban centers also provide a greater number of social opportunities, which substantiate Michael Batty's observation that "information flows both replace and complement material flows of resources" in urban contexts (2008, p. 770). The increase of population in urban areas enabled people to avail themselves of a larger variety of groups, including those with selective membership (such as co-ethnic groups, religious groups, and sporting and spectator groups) as well as groups based on the happenstance of mere proximity such as living in the same structure or neighborhood. Urban centers have many points of contact that result in what Granovetter (1973) has identified as "weak ties" of social engagement that enable low-cost casual interactions to be productively transformed into information leading to desired outcomes such as assistance, employment, and resource acquisition.

Weak-tie relationships also may result in the perception of increased security in cities through "safety in numbers" (Cowgill, 2003, p. 47). Although this safety generally is assumed to be related to military action, factors of increased group size likely resulted in other types of risk reduction. In the animal kingdom, "safety in numbers" has a counterintuitive beneficial effect related to food intake because group size can increase foraging efficiency as noted by Winterhalder et al. (1999, pp. 324–325). This effect is found in a number of group-living organisms ranging from birds to carnivores, and can be conceptualized for human groups as well. For small-scale and impoverished individuals, perhaps the city with its masses of population is equivalent to a herd of wildebeest or school of fish, in which the sheer mass of population enables each individual to weather risk on the personal level more easily (bad things might happen to someone else, because there were so many other people around). Individuals also had a greater range of consumption opportunities, because cities draw in a large supply of goods from the hinterlands despite the costs of transportation (cf. Drennan, 1984). The resultant diversity and abundance of provisions provide the opportunity for sanctioned acquisition as well as what Blurton Jones (1987, pp. 31–35) and others have described for forager groups as "tolerated theft" or "tolerated scrounging."

The mutualisms expressed within urban settings can be assessed through a variety of theoretical frameworks, one of the most promising of which is niche construction theory. First developed in biology to explain the evolutionary advantages of species mutualisms, niche-construction theory has recently been utilized by archaeologists to analyze the way in which organisms and their environments act upon each other in reciprocal and dynamic relationships (Crawford, 2014, p. 147; Laland & O'Brien, 2011, p. 193; B. Smith, 2011; Spisak, O'Brien, Nicholson, & Van Vugt, 2015). Mutualisms are not limited to physical interactions, however, as humans also learn from and integrate their activities with other human beings through learning and communication, creating interpersonal configurations described by Laland and O'Brien (2011, p. 198) as *social niche construction*. The effects of social niche construction are particularly visible in urban areas, where people interact with and modify their social ecosystems through many forms of interdependence (cf. Spisak et al., 2015). Specialized craftsmakers became dependent on a larger consumer base that in turn made use of diverse goods for social signaling; elites and their followers relied upon each other in cycles of patronage and service; and manufacturers, workers, and consumers created interdependencies of production and distribution.

One of the most significant urban social niches is found in the creation of institutions as collective groups charged with particular tasks related to newly emergent economic, social, political, and ritual needs. The concept of the "institution" merits some unpacking in order to understand its impact on the urban form, as the term is utilized to describe both physical and conceptual forms of interaction. Institutions can be framed as a codification of culture, consisting of "organizational and ideological norms that shape practical interactions of agents and communities" (Frachetti, 2012, p. 5, following North, 1990; see also Tomasello, 1999, p. 510). But institutions also exist as physical entities that have a visual, material presence. In the modern world, examples of such "institutions" include the armed forces, schools and colleges, religious venues, police precincts, halfway houses, hospitals, prisons, and courthouses. Many of these configurations were historically developed for the very first time in cities, when leaders at a variety of scales materialized norms into configurations whose physical presence resulted in fixed-place recognition of their authority. Institutions are *how* states and cities grow up together and represent the way in which the political entity of the state is dependent on cities for growth and viability.

Anthropologists have long identified the importance of institutions in the creation of social complexity, citing institutions as a key factor in the transition from a kin-based society to hierarchies that functioned through impersonal "offices" that could be held by different individuals interchangeably (e.g., Flannery, 1972; Fried, 1967; Spencer, 2009). Institutions enable programs of action to be initiated and developed through teamwork directed by replaceable bureaucrats rather than being dependent on the charisma of a single individual. Institutions are designed to outlast their founders and constitute entities to which large numbers of differentially enfranchised individuals can belong. Within institutions, leadership provides a kind of "safety in numbers" that buffers individuals from threats and abuses while often providing structured opportunities for training and skill upgrades (Spisak et al., 2015). In urban centers, institutions such as guilds and workshops provided employment opportunities in which labor was compensated in cash or rations, while non-governmental organizations provided food (and sometimes, shelter) in return for adherence to doctrine. By far, the most significant institutions in cities, however, were those formed by government organizations to address needs of urban spatial organization, infrastructure, and defense.

When institutions grow in size, their risk-buffering capacity increases proportionately, such that "the greater the degree of risk, the wider the range of social institutions to share the risk" (Browman, 1987, p. 173). Designed to outlive any single participant, institutions nonetheless have the same risk-management strategies as individuals in focusing on immediate and short-term gain. Even when long-term solutions are sought, managers are constrained and have more intense and more diverse pressures on investment and performance (Blaikie & Brookfield, 1987, p. 10).[2] Institutions articulate short-term gains through a variety of physical means including resource assessment and land-modification projects. Like modern ones, ancient cities were invested with formal infrastructure signaling the presence of an organizational dynamic beyond the level of the household, including roads (Branting, 2004), sewer and drainage systems (Ortloff & Kolata, 1989), water provisioning (Hodge, 2013), open spaces (Fleischer, 2014), internal walls (Kenoyer, 1998), fortifications and gateways (Mohanty & Smith, 2008), harborworks (Marriner & Morhange, 2007), marketplaces (Hutson, 2000), and other venues that were the simultaneous result of top-down managerial initiatives and bottom-up individual actions. The cumulative effects of landscape-scale interactions can be described as "landesque capital," defined as the materialization of large-scale labor investment (Blaikie & Brookfield, 1987, p. 9) that results in a cumulative effect over generations (Erickson & Walker, 2009).

Urban institutional investment in landesque capital extended to the surrounding countryside. Cities and their rural hinterlands are mutually constituted and interdependent (Adams, 1972; M. L. Smith, 2014; Yoffee, 1995, p. 284), but the relationship is not equal because cities have decision-enforcing capacity and concentrated resources for investment as they draw in raw materials, food, and labor for the support of urban populations (Blaikie & Brookfield, 1987, p. 21; Pires, 2004). Leaving aside the question of whether hierarchical organization is required for landscape-scale changes such as irrigation, it is clear that the construction and maintenance of terraces, canals, and other forms of productive agricultural infrastructure necessitate the coordination of large numbers of people. Archaeological indicators of ancient urban-dominated infrastructure can be seen, for example, in the agricultural terraces in the hinterlands of Cerro Jazmín in Mesoamerica (Peréz Rodríguez, Anderson, & Neff, 2011), the raised fields of the highland Andes (Erickson & Walker, 2009), and Greek and Roman harbors in the Mediterranean (Marriner & Morhange, 2007). The economic pull of ancient cities also restructured distant supply lines for daily-use goods in indirect ways (see, e.g., Larsen, 1987, p. 51 for the woolen textile trade of the second-millennium BC Mesopotamian cities and Dermody et al., 2014 for the Roman practice of importing grain from North Africa as an agricultural outsourcing project that bypassed local Italian farmers). The resulting structural adjustments for the benefit of urban populations were not unlike the conditions that prevail in the global trade of foodstuffs and raw materials today (see Adger, 2000; Lambin et al., 2001; Logan, 2012; Wutich & Brewis, 2014).

Although cities hold the balance of power in urban–rural relationships, rural residents also benefit socially and materially from the presence of urban centers. Familiarity with urban styles enables rural residents to demonstrate savoir-faire and other forms of cultural capital (Jeffery, Jeffery, & Jeffery, 2011), as well as acquiring remittances that can be invested in hinterland settings. Through circular migration for employment and education, rural inhabitants increase their capacity for "foraging" success by diversifying their activities even if they are tied to the land, enabling them to increase household wealth and resilience. Researchers have noted in studies of modern dwellers of the Himalayan region that urban migration and remittances can enable rural households to recover from an environmental shock but that the practice is most effective when utilized as a consistent strategy of diversification and not just as a response to a specific event (Gioli, Khan, Bisht, & Scheffran, 2014). Households that engaged in migration prior to environmental shortfall used remittances to engage in longer-term investments such as education and health that helped to weather shocks when they did occur, whereas households who initiated migration only after adversity found themselves using remittances for basic survival needs in a manner that was primarily reactive. The reliance on rural dwellers upon cities illustrates that the interactions created in the process of urban formation result in dynamic, mutually reinforcing configurations in both the urban and hinterland environments.

DISCUSSION

A longitudinal assessment of the transition from foraging to settled societies indicates that local fluctuations of resource productivity were likely to have been recognized by our earliest human ancestors. Approaches to subsistence throughout human history, however, have been based on an expectation of individualized mitigation of environmental challenges: through self-provisioning in forager societies, through household storage in agricultural societies and through networking and alliances in urban societies. Although cities were the places where individuals might have been expected to express the greatest anxiety about access to resources, urban centers actually provided the greatest stability from the individual perspective because of the comparatively greater array of foraging opportunities made available through mutually reinforcing phenomena that promoted urban growth (Figs. 1 and 2).

Batty (2008, p. 769) notes that cities "appear to be barely sustainable but paradoxically resilient networks," an observation confirmed by the fact that many ancient cities survived for centuries and weathered numerous environmental and political challenges. Even in the world's first cities, resource-acquisition anxiety was counterbalanced by the coincident growth of governmental and non-governmental institutions. Temples, guilds, armies, and other collectivities became enmeshed in the social conditions of its members under conditions of either normal function (cf. Van de Pol & Kuijpers, 2005) or under conditions of crisis (Richardson & McBride,

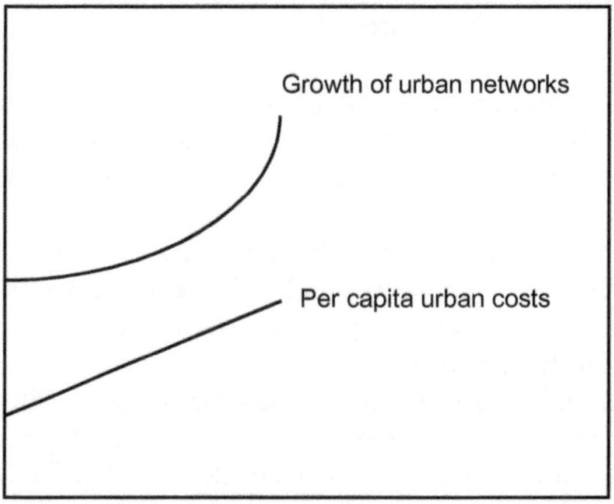

Fig. 1. From the perspective of any individual, it almost always makes sense to come into the city, because individuals can immediately utilize the networks of which they are now a part to provide a more diverse set of solutions to the problem of obtaining food (or other desirables ranging from consumer goods to finding mates). For each new person who comes into the city, there is a single-unit incremental increase in costs (of food, raw materials, water, and other inputs) that produces a linear trajectory, but network construction is an exponential function (following Metcalfe's Law, which states that "the value of a network grows as the square of the number of its users," Metcalfe, 2013, p. 26).

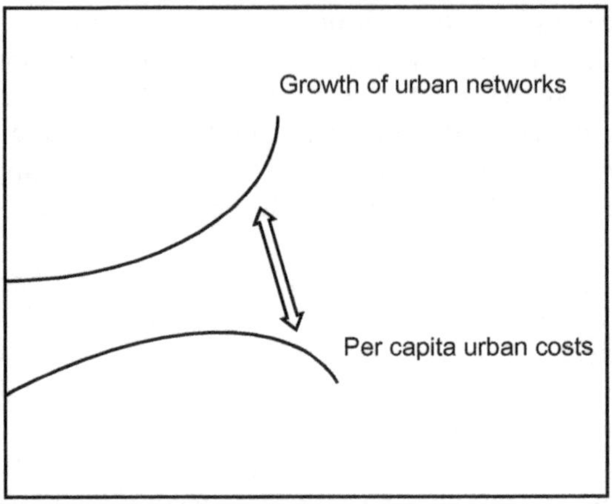

Fig. 2. From the institutional perspective, costs per individual increase initially because of the need to build transportation and storage infrastructure, but then the cost per person decreases as more people come into the urban environment in which that infrastructure is already built (cf. O'Meara, 1999). As a result, cities become larger and larger.

2009). These institutions provided conditions for what could be termed "optimal social foraging" as long as institutions and infrastructure were maintained at a functional level. Cases of modern cities surviving, and even growing, under conditions of warfare and the bankruptcy of formal governments indicate the extent to which a city *sui generis* represents a locus of sustainability from the individual and household perspective, even if the goal of sustainability is never explicitly articulated by any individuals or institutions within the urban realm.[3]

When, then, did "sustainability" become a deliberate concept? The answer is: only recently, when the coincidence of real human effects on the population was combined with measuring devices that enable the quantification of change. The current level of global population is a unique configuration compared to prior eras; as identified by Klein Goldewijk, Beusen, and Janssen (2010, pp. 568–570), global populations have risen dramatically over time from a total of only six million people on Earth in 10,000 BC, to 18 million by 5,000 BC at the time of the earliest cities, to 200 million in AD 500. Global population has been rising extremely rapidly in the modern era, with a total count of 3.6 billion people in 1970 that has nearly doubled today. Not only is overall global population density higher than at any time in human history, but humans also are experiencing population density in ways that are distinct from ancestral configurations of group size and mobility.

Our current array of measuring devices is unprecedented in human history, ranging from long-term climate records captured in ice cores to real-time satellite imagery that can monitor vegetation growth or oceanic surface temperatures over large expanses of the earth. Many parts of the world now have precise temperature and rainfall records for the past century or more, providing robust data sets from which can be gleaned statistical evaluations of past fluctuations that also provide a baseline for projecting future conditions. These measurements, along with the simultaneous development of human technologies capable of altering the environment on a global scale, have resulted in the conceptualization of the modern world as a "risk society" (Beck, 1992[1986]), in which advances in mathematics and quantification provide the basis for the calculation of risk (White, 2010, p. 91).

Because resource acquisition problems are socially defined, the facts of population growth and the quantification of risk do not in and of themselves necessarily result in a perception of a need for human–environmental rebalancing. The next logical question to ask is *why* sustainability has become a focal concept today. Three potential explanations can be identified: a philosophical turn against modernism and consumption; an interest in using environmental language in the pursuit of social justice; and an expression of anxiety by the middle class.

The philosophical assessment of sustainability can be placed within a long-running theme on the evils of the modern consumer society (for a recent summary and contextualization of this theme, see Pallandini-Simányi, 2014). With a particular reference to the natural environment, Hames (2007) has observed that the myth of the "noble savage" with its connotations of premodern environmental stewardship has a long history in the Euro-American philosophical tradition. Time and again, archaeological research has indicated the ways in which ancient societies outstripped their resources resulting in long-term damage (see, e.g., Redman, 1999), but popular media continues to engage with the trope that small-scale, indigenous societies harmonize with nature and thus constitute a more authentic form of human–environmental interaction.

Sustainability also has become part of the language of environmental and social justice (Harper, 2007). In analyzing the goals of sustainability science, Redman (2014, p. 3) notes that sustainability is "aspirational and requires iterations of improvements," and that ideals of advocacy and activism permeate the language of environmental planning. The concept of "future discounting" can be applied to these social movements as well: physical outcomes are encompassed within so many complexities that the effects of mitigation are largely unknowable (cf. Weitzman, 1998), but a focus on social justice under the rubric of "sustainability" still yields benefits in the form of measurable community cohesion and social well-being in the present.

A final rationale for the modern concern with "sustainability" is an expression of anxieties sustained by a particular type of group within the modern context, namely, the middle class. Although there is good evidence for an intermediate producer–consumer stratum that emerged in tandem with the world's first cities, the numbers and proportion of the middle class today are much larger than in any prior era. Concern for environmental sustainability as an expression of being able to pass along inheritance to heirs may be a psychological outgrowth of a generalized state of anxiety about a challenging, risk-laden future. In India, for example, Amita Baviskar has discussed the zeal of urban "bourgeois environmentalists" for whom "environmentalism is a mode of expressing and addressing their anxieties about themselves in relation to their habitat, that is, their physical surroundings, both proximate and distant, and other species" (2011, p. 401; see also Harper, 2007 for Hungary; Marcuse, 1998; Strang, 2011 for Australia). Although risk and vulnerability are concepts familiar to all social strata, the concept of sustainability may well be a middle-class one indicative of a desire to protect limited assets in coordination with the apparatus of the state (e.g., through legislation protecting the environment, class-action lawsuits requiring the mitigation of damaged ecosystems, and fund-raising for direct acquisition and protection of targeted natural areas).

CONCLUSIONS

Humans have entered the current age of high population density with a cognitive and social apparatus that is more comfortably attuned to the perception and mitigation of short-term risks (such as natural disasters, warfare, interpersonal violence, and household dissolution) than to the long-term risks encompassed in incremental climate change, sea-level rise, species extinction and global resource depletion. From the perspective of the ancient individual, the acquisition of desired resources was not a matter of sustainability viewed as long-term stasis, but survival as a matter of short-term access. Just at the historical moment when population pressure might have induced a greater level of anxiety, however, cities emerged as a complex nexus of provisioning opportunities. Through the development of redistributive institutions, markets, and other resource-providing mechanisms, individuals could revert to age-old foraging strategies amidst a vastly expanded array of resource types and localities. With their far-flung economic networks and capacity for political leverage over hinterlands, cities enabled environmental challenges to be masked by an apparent abundance. Even in the rare cases when long-term measurements were developed, for example, in the form of the Nilometer that measured the annual height of Egypt's life-giving river, the concern was for the mitigation of proximate shortfalls or, at the most, the calculation of the probability of the type, periodicity, and scale of shortfalls that could be expected.

Although the concept of sustainability now appears essential for long-term human viability and the future of our planet, it is a concept that faces considerable challenges in both articulation and actualization. Individuals, households, and institutions have identified stop-gap strategies over time and have significant difficulties in cognizing long-term risks, meaning that "sustainability" as a future-directed stability optimum is an unfamiliar and recent paradigm. The benchmark for sustainability is unclear, and each permutation is equally valid: whether to strive for some past pristine condition, a reduction of present impact, or a target of some percentage of future calculated growth. Although measurement devices can pinpoint with greater and greater accuracy the historical variability of climate and the cause-and-effect relationships of human actions on the environment, those devices cannot in and of themselves identify the best approach for present-day and future actions. Given that human cognition is most effectively directed to the understanding of short-term consequences, concerns for long-term "sustainability" must be taught rather than assumed as an inherent perceptive capacity.

NOTES

1. Given its prevalence in our species' behavior for the past two million years, foraging is a highly successful adaptation that should be considered the default adaptive mode. This mode of resource acquisition is highly distinct from sedentary agriculture, which only became practiced (or practicable, see Richerson et al., 2001) after the Holocene. Sedentary agriculture would appear to greatly reduce mobility because of the necessity to continually tend crops throughout the growing cycle and safeguard the post-harvest comestibles, but even sedentary agriculturalists still

engage in strategies of flexibility that mimic forager mobility (see Winterhalder & Goland, 1997). These strategies include diversification of crop types or herd composition, sharing of produce, and storage of temporary surpluses for future use (Winterhalder, Lu, & Tucker, 1999, pp. 335–336); innovation by changing crop mixtures under conditions of stress (M. L. Smith, 2006); taking up new economic activities such as pastoralism (Murty & Sontheimer, 1981); modifying cultural norms to engage in practices such as early marriage and adoption of offspring (Forbes, 1989); sending some family members for external work with the goal of remittances (Adger, 2000; Whittaker & Goody, 2001); and/or going into debt (Wutich & Brewis, 2014).

2. Institutions can, of course, undertake very significant long-term planning for the future, as seen in the urban water-resource projects of New York City starting in the 19th century (Pires, 2004) and Los Angeles in the early 20th century (Gumprecht, 2005). Both of these examples illustrate that the planning for the future still contains a strong present-gain component, however, as seen in the explicit statements of politicians of their power over hinterland resources.

3. A related question, beyond the scope of this paper: if cities are loci of sustainability in and of themselves, why do cities ever collapse? One telling point is the way in which ancient cities collapsed but modern ones rarely do; if anything, as states become unstable their urban areas tend to be viewed as the only viable node of interaction and continue to grow (examples of which include Kinshasa and other cities in so-called "failed states"; see Iyenda, 2005). Although ancient states had fewer means of direct control available to them in a pre-firearms era, it is not clear the extent to which the strength or weakness of a political authority could affect a city if that city was a robust node of ritual or economic activity.

ACKNOWLEDGMENTS

Many thanks to Evan Carlson, Sarah Chandlee, Ashley Fent, Aditi Halbe, Brittany Jackson, Jillian Jones, Kanika Kalra, and Trevor Van Damme for sharing conversations about urbanism and the environment at UCLA; in addition, thanks to Melinda Zeder for observations about biological diversity in cities. I very much appreciate the extensive and helpful comments from two anonymous reviewers as well as the support for the research expressed by Donald Wood.

REFERENCES

Adams, R. M. (1972). Patterns of urbanization in early Southern Mesopotamia. In P. J. Ucko, R. Tringham & G. W. Dimbleby (Eds.), *Man, settlement and urbanism* (pp. 735–749). Cambridge, MA: Schenkman.

Adger, W. N. (2000). Social and ecological resilience: Are they related? *Progress in Human Geography*, *24*, 347–364.

Anscheutz, K. F., Wilshusen, R. H., & Scheick, C. L. (2001). An archaeology of landscapes: Perspectives and directions. *Journal of Archaeological Research*, *9*, 157–211.

Barton, C. M., Ullah, I. I. T., Bergin, S. M., Mitasova, H., & Sarjoughian, H. (2012). Looking for the future in the past: Long-term change in socioecological systems. *Ecological Modeling*, *241*, 42–53.

Batty, M. (2008). The size, scale, and shape of cities. *Science*, *319*, 769–771.

Baviskar, A. (2011). Cows, cars and cycle-rickshaws: Bourgeois environmentalists and the battle for Delhi's streets. In A. Baviskar & R. Ray (Eds.), *Elite and everyman: The cultural politics of the Indian middle classes* (pp. 391–418). New Delhi: Routledge.

Beck, U. (1992[1986]). *Risk society: Towards a new modernity*. Los Angeles, CA: Sage.

Bell, B. (1970). The oldest records of the Nile floods. *The Geographical Journal*, *136*, 569–573.

Blaikie, P., & Brookfield, H. (1987). Defining and debating the problem. In P. Blaikie & H. Brookfield (Eds.), *Land degradation and society* (pp. 1–26). London: Methuen.

Bliege Bird, R., Bird, D. W., Codding, B. F., Parker, C. H., & Jones, J. H. (2008). The "fire stick farming" hypothesis: Australian Aboriginal foraging strategies, biodiversity, and anthropogenic fire mosaics. *Proceedings of the National Academy of Sciences*, *105*, 14796–14801.

Blurton Jones, N. G. (1987). Tolerated theft, suggestions about the ecology and evolution of sharing, hoarding and scrounging. *Social Science Information*, *26*, 31–54.

Bogaard, A., Charles, M., Twiss, K. C., Fairbairn, A., Yalman, N., Filipović, D., Arzu Demirergi, G., Ertuğ, F., Russell, N., & Henecke, J. (2009). Private pantries and celebrated surplus: Storing and sharing food at Neolithic Catalhöyük, central Anatolia. *Antiquity*, *83*, 649–668.

Borejsza, A., Rodríguez López, I., Frederick, C. D., & Bateman, M. D. (2008). Agricultural slope management and soil erosion at La Laguna, Tlaxcala, Mexico. *Journal of Archaeological Science*, *35*, 1854–1866.

Boyd, R., Richerson, P. J., & Henrich, J. (2013). Cultural evolution of technology: Facts and theories. In P. J. Richerson & M. H. Christiansen (Eds.), Cultural evolution: Society, technology, language, and religion (pp. 119–142). Cambridge, MA: The MIT Press.

Branting, S. A. (2004). *Iron Age pedestrians at Kerkenes Dağ: An archaeological GIS-T approach to movement and transportation*. Ph.D. dissertation, State University of New York, Buffalo.

Browman, D. L. (1987). Agro-pastoral risk management in the central Andes. *Research in Economic Anthropology*, *8*, 171–200.

Carballo, D. M., Roscoe, P., & Feinman, G. M. (2014). Cooperation and collective action in the cultural evolution of complex societies. *Journal of Archaeological Method and Theory*, *21*, 98–133.

Colten, C. E. (2005). *An unnatural metropolis: Wresting New Orleans from nature*. Baton Rouge, LA: Louisiana State University Press.

Costanza, R., van der Leeuw, S., Hibbard, K., Aulenbach, S., Brewer, S., Burek, M., ... Steffen, W. (2012). Developing an Integrated History and Future of People on Earth (IHOPE). *Current Opinion in Environmental Sustainability*, *4*, 106–114.

Cowgill, G. L. (2003). Teotihuacan: Cosmic glories and mundane needs. In M. L. Smith (Ed.), *The social construction of ancient cities* (pp. 37–55). Washington, DC: Smithsonian Institution Press.

Crawford, G. W. (2014). *Food production and niche construction in pre-contact southern Ontario*. Midwest Archaeological Conference Inc. Occasional Papers No. 1, pp. 135–160.

Dermody, B. J., van Beek, R. P. H., Meeks, E., Klein Goldewijk, K., Schiedel, W., van der Velde, Y., Bierkens, M. F. P., Wassen, M. J., & Dekker, S. C. (2014). A virtual water network of the Roman world. *Hydrology and Earth Systems Sciences*, *18*, 5025–5040.

di Lernia, S. (2006). Building monuments, creating identity: Cattle cult as a social response to rapid environmental changes in the Holocene Sahara. *Quaternary International*, *151*, 50–62.

Douglas, I., Kurshid, A., Maghenda, M., McDonnell, Y., McLean, L., & Campbell, J. (2008). Unjust waters: Climate change, flooding and the urban poor in Africa. *Environment and Urbanization*, *20*, 187–205.

Drennan, R. D. (1984). Long-distance transport costs in pre-Hispanic Mesoamerica. *American Anthropologist*, *86*, 105–112.

Dubas, K. M., & Jonsson, P. (2005). Rationality in consumer decision making. *Proceedings of the Academy of Marketing Studies*, *10*, 5–8.

Ellis, E. C., Kaplan, J. O., Fuller, D. Q., Vavrus, S., Goldewijk, K. K., & Verburg, P. H. (2013). Used planet: A global history. *Proceedings of the National Academy of Sciences*, *110*, 7978–7985.

Erickson, C. L., & Walker, C. L. (2009). Precolumbian causeways and canals as landesque capital. In J. E. Snead, C. L. Erickson & J. A. Darling (Eds.), *Landscapes of movement: Trails, paths, and roads in anthropological perspective* (pp. 233–252). Philadelphia, PA: University of Pennsylvania Museum of Archaeology and Anthropology.

Flannery, K. V. (1972). The cultural evolution of civilizations. *Annual Review of Ecology and Systematics*, *3*, 399–426.

Fleischer, J. (2014). The complexity of public space at the Swahili town of Songo Mnara, Tanzania. *Journal of Anthropological Archaeology*, *35*, 1–22.

Forbes, H. (1989). Of grandfathers and grand theories: The hierarchised ordering of responses to hazard in a Greek rural community. In P. Halstead & J. M. O'Shea (Eds.), *Bad year economics: Cultural responses to risk and uncertainty* (pp. 87–97). Cambridge, MA: Cambridge University Press.

Frachetti, M. D. (2012). Multiregional emergence of mobile pastoralism and nonuniform institutional complexity across Eurasia. *Current Anthropology*, *53*, 2–38.

Fried, M. (1967). *The evolution of political society*. New York, NY: Random House.

Fuller, D. Q., Van Etten, J., Manning, K., Castillo, C., Kingwell-Banham, E., Weisskopf, A., … Hijmans, R. J. (2011). The contribution of rice agriculture and livestock pastoralism to prehistoric methane levels: An archaeological assessment. *The Holocene SI*, 1–17.

Gioli, G., Khan, T., Bisht, S., & Scheffran, J. (2014). Migration as an adaptation strategy and its gendered implications: A case study from the upper Indus basin. *Mountain Research and Development*, *34*, 255–265.

Granovetter, M. S. (1973). The strength of weak ties. *American Journal of Sociology*, *78*, 1360–1380.

Gumprecht, B. (2005). Who killed the Los Angeles river? In W. Deverell & G. Hise (Eds.), *Land of sunshine: An environmental history of metropolitan Los Angeles* (pp. 115–134). Pittsburgh, PA: University of Pittsburgh Press.

Halstead, P., & O'Shea, J. (1989). Introduction: Cultural responses to risk and uncertainty. In P. Halstead & J. O'Shea (Eds.), *Bad year economics: Cultural responses to risk and uncertainty* (pp. 1–7). Cambridge, MA: Cambridge University Press.

Hames, R. (2007). The ecologically noble savage debate. *Annual Review of Anthropology*, *36*, 177–190.

Harper, K. (2007). Does everyone suffer alike? Race, class, and place in Hungarian environmentalism. In C. A. Maida (Ed.), *Sustainability and communities of place* (pp. 109–126). New York, NY: Berghahn.

Hegmon, M. (1991). The risks of sharing and sharing as risk reduction: Interhousehold food sharing in egalitarian societies. In S. A. Gregg (Ed.), *Between bands and states* (pp. 309–329). Carbondale, IL: Southern Illinois University Center for Archaeological Investigations Occasional Paper No. 9.

Hodge, A. T. (2013). Aqueducts and water supply. In J. DeRose Evans (Ed.), *A companion to the archaeology of the Roman Republic* (pp. 285–295). Oxford: Blackwell.

Hutson, S. R. (2000). Carnival and contestation in the Aztec marketplace. *Dialectical Archaeology*, *25*, 123–149.

Iyenda, G. (2005). Street enterprises, urban livelihoods, and poverty in Kinshasa. *Environment and Urbanization*, *17*, 55–67.

Jeffery, R., Jeffery, P., & Jeffery, C. (2011). Are rich rural jats middle-class? In A. Baviskar & R. Ray (Eds.), *Elite and everyman: The cultural politics of the Indian middle classes* (pp. 140–163). New Delhi: Routledge.

Kates, R. W., Clark, W. C., Corell, R., Hall, J. M., Jaeger, C. C., Lowe, I., … Svedin, U. (2001). *Sustainability science (December 2000)*. KSG Working Paper No. RWP00-018. Retrieved from http://dx.doi.org/10.2139/ssrn.257359

Kemper, R. V. (2010). The extended community: Migration and transformation in Tzintzuntzan, Mexico. In G. Gmelch, R. V. Kemper & W. P. Zenner (Eds.), *Urban Life* (pp. 285–299). Long Grove, IL: Waveland Press.

Kenoyer, J. M. (1998). *Ancient cities of the Indus Valley civilization*. Karachi: Oxford University Press.

Kintigh, K. W., Altschul, J. H., Beaudry, M. C., Drennan, R. D., Kinzig, A. P., Kohler, T. A., … Zeder, M. A. (2014). Grand challenges for archaeology. *American Antiquity*, *79*, 5–24.

Klein Goldewijk, K., Beusen, A., & Janssen, P. (2010). Long-term dynamic modeling of global population and built-up area in a spatially explicit way: HYDE 3.1. *The Holocene*, *20*, 565–573.

Laland, K. N., & O'Brien, M. J. (2011). Cultural niche construction: An introduction. *Biological Theory*, *6*, 191–202.

Lambin, E. F., Turner, B. L., Geist, H. J., Agbola, S. B., Angelsen, A., Bruce, J. W., … Xu, J. (2001). The causes of land-use and land-cover change: Moving beyond the myths. *Global Environmental Change*, *11*, 261–269.

Larsen, M. T. (1987). Commercial networks in the ancient Near East. In M. Rowlands, M. Larsen & K. Kristiansen (Eds.), *Centre and periphery in the ancient world* (pp. 47–56). Cambridge, MA: Cambridge University Press.

Layton, R., O'Hara, S., & Bilsborough, A. (2012). Antiquity and social functions of multilevel social organization among human hunter-gatherers. *International Journal of Primatology*, *33*, 1215–1245.

Loehle, C., & Eschenbach, W. (2011). Historical bird and terrestrial mammal extinction rates and causes. *Diversity and Distributions, 18*, 1–8.

Logan, A. L. (2012). *A history of food without history: Food, trade, and environment in west-central Ghana in the second millennium* AD. Ph.D. dissertation, Department of Anthropology, University of Michigan.

Marcuse, P. (1998). Sustainability is not enough. *Environment and Urbanization, 10*, 103–111.

Marriner, N., & Morhange, C. (2007). Geoscience of ancient Mediterranean harbours. *Earth-Science Reviews, 80*, 137–194.

Metcalfe, B. (2013). Metcalfe's law after 40 years of Ethernet. *Computer, 46*(12), 26–31.

Minnis, P. (2010). Ancient anthropogenic ecology: Beyond the past. In R. M. Dean (Ed.), *The Archaeology of Anthropogenic Environments* (pp. 373–379). Carbondale, IL: Center for Archaeological Investigations Southern Illinois University. Carbondale Occasional Paper No. 37.

Mohanty, R. K., & Smith, M. L. (2008). *Excavations at Sisupalgarh.* New Delhi: Indian Archaeological Society.

Murty, M. L. K., & Sontheimer, G. D. (1981). Prehistoric background to pastoralism in the Southern Deccan in the light of oral tradition and cults of some pastoral communities. *Anthropos, 75*, 163–184.

Mustafa, D. (2005). The production of an urban hazardscape in Pakistan: Modernity, vulnerability, and the range of choice. *Annals of the Association of American Geographers, 95*, 566–586.

North, D. C. (1990). *Institutions, institutional change and economic performance.* Cambridge, MA: Cambridge University Press.

O'Meara, M. (1999). *Reinventing cities for people and the planet.* Washington, DC: Worldwatch Institute Worldwatch Paper No. 147.

Ortloff, C. R., & Kolata, A. L. (1989). Hydraulic analysis of Tiwanaku aqueduct structures at Lukurmata and Pajchiri, Bolivia. *Journal of Archaeological Science, 16*, 513–535.

Owens, S. (2003). Is there a meaningful definition of sustainability? *Plant Genetic Resources, 1*, 5–9.

Pallandini-Simányi, L. (2014). *Consumption norms and everyday ethics.* New York, NY: Palgrave Macmillan.

Peréz Rodríguez, V., Anderson, K. C., & Neff, M. K. (2011). The Cerro Jazmín archaeological project: Investigating prehispanic urbanism and its environmental impact in the Mixteca Alta, Oaxaca, Mexico. *Journal of Field Archaeology, 36*(2), 83–99.

Pires, M. (2004). Watershed protection for a world city: The case of New York. *Land Use Policy, 21*, 161–175.

Prado, J. L., Arroyo-Cabrales, J., Johnson, E., Alberdi, M. T., & Polaco, O. J. (2012). New World proboscidean extinctions: Comparisons between North and South America. *Archaeological and Anthropological Sciences*, doi 10.1007/s12520-012-0094-3.

Rautman, A. E. (2013). Social integration and the built environment of aggregated communities in the North American Puebloan Southwest. In J. Birch (Ed.), *From prehistoric villages to cities: Settlement aggregation and community transformation* (pp. 111–133). New York, NY: Routledge.

Redman, C. L. (1999). *Human impact on ancient environments.* Tucson, AZ: University of Arizona Press.

Redman, C. L. (2014). Should sustainability and resilience be combined or remain distinct pursuits? *Ecology and Society, 19*, 37. Retrieved from http://dx.doi.org/10.5751/ES-06390-190237

Richardson, G., & McBride, M. (2009). Religion, longevity, and cooperation: The case of the craft guild. *Journal of Economic Behavior and Organization*, *71*, 172–186.

Richerson, P. J., Boyd, R., & Bettinger, R. L. (2001). Was agriculture impossible during the pleistocene but mandatory during the Holocene? A climate change hypothesis. *American Antiquity*, *66*, 387–411.

Rise, J., Åstrøm, A. N., & Sutton, S. (1998). Predicting intentions and use of dental floss among adolescents: An application of the theory of planned behavior. *Psychology and Health*, *13*, 223–236.

Roos, C. L., Sullivan III, A. P., & McNamee, C. (2010). Paleoecological evidence for systematic indigeous burning in the upland Southwest. In R. M. Dean (Ed.), *The archaeology of anthropogenic environments* (pp. 142–171). Carbondale, IL: Center for Archaeological Investigations Southern Illinois University. Carbondale Occasional Paper No. 37.

Rosman, K. J. R., Chisholm, W., Hong, S., Candelone, J.-P., & Boutron, C. F. (1997). Lead from Carthaginian and Roman Spanish mines isotopically identified in Greenland ice dated from 600 B.C. to 300 AD. *Environmental Science and Technology*, *31*, 3413–3416.

Sanders, J., Nee, V., & Sernau, S. (2002). Asian immigrants' reliance on social ties in a multi-ethnic labor market. *Social Forces*, *81*, 281–314.

Smith, B. D. (2001). Low-level food production. *Journal of Archaeological Research*, *9*, 1–43.

Smith, B. D. (2011). A cultural niche construction theory of initial domestication. *Biological Theory*, *6*, 260–271.

Smith, M. E. (2010). Sprawl, squatters and sustainable cities: Can archaeological data shed light on modern urban issues? *Cambridge Archaeological Journal*, *20*, 229–253.

Smith, M. L. (2006). How ancient agriculturalists managed yield fluctuations through crop selection and reliance on wild plants: An example from central India. *Economic Botany*, *60*, 39–48.

Smith, M. L. (2010). *A prehistory of ordinary people*. Tucson, AZ: University of Arizona Press.

Smith, M. L. (2013). Caste as a cooperative economic entitlement strategy in complex societies of the Indian subcontinent and subsaharan Africa. In D. M. Carballo (Ed.), *Cooperation and collective action* (pp. 275–297). Boulder, CO: University Press of Colorado.

Smith, M. L. (2014). The archaeology of urban landscapes. *Annual Review of Anthropology*, *43*, 307–323.

Spencer, C. S. (2009). Testing the morphogenesist model of primary state formation: The Zapotec case. In A. M. Prentiss, I. Kuijt & J. C. Chatters (Eds.), *Macroevolution in human prehistory* (pp. 133–155). New York, NY: Springer.

Spisak, B. R., O'Brien, M. J., Nicholson, N., & Van Vugt, M. (2015). Niche construction and the evolution of leadership. *Academy of Management Review*, *40*, 291–306.

Strang, V. (2011). Fluid forms: Owning water in Australia. In V. Strang & M. Busse (Eds.), *Ownership and appropriation* (pp. 171–195). Oxford: Berg.

Tomasello, M. (1999). The human adaptation for culture. *Annual Review of Anthropology*, *28*, 509–529.

Van de Pol, L., & Kuijpers, E. (2005). Poor women's migration to the city: The attraction of Amsterdam health care and social assistance in early modern times. *Journal of Urban History*, *32*, 44–60.

Weitzman, M. L. (1998). Why the far-distant future should be discounted at its lowest possible rate. *Journal of Environmental Economics and Management*, *36*, 201–208.

White, I. (2010). *Water and the city: Risk, resilience and planning for a sustainable future*. New York, NY: Routledge.

Whittaker, D., & Goody, J. (2001). Rural manufacturing in the Rouergue from antiquity to the present: The examples of pottery and cheese. *Comparative Studies in Society and History*, *43*(2), 225–245.

Winterhalder, B., & Goland, C. (1997). An evolutionary ecology perspective on diet choice, risk, and plant domestication. In K. J. Gremillion (Ed.), *People, plants, and landscape: Studies in paleoethnobotany* (pp. 123–160). Tuscaloosa, AL: The University of Alabama Press.

Winterhalder, B., Lu, F., & Tucker, B. (1999). Risk-sensitive adaptive tactics: Models and evidence from subsistence studies in biology and anthropology. *Journal of Archaeological Research*, *7*, 301–348.

Wutich, A., & Brewis, A. (2014). Food, water, and scarcity: Toward a broader anthropology of resource insecurity. *Current Anthropology*, *55*, 444–468.

Yoffee, N. (1995). Political economy in early Mesopotamian states. *Annual Review of Anthropology*, *24*, 281–311.

Zeder, M. A. (2012). The broad spectrum revolution at 40: Resource diversity, intensification, and an alternative to optimal foraging explanations. *Journal of Anthropological Archaeology*, *31*, 241–264.

MERCHANT RESILIENCY AND CLIMATE HAZARD VULNERABILITY IN THE URBAN PHILIPPINES: ANTHROPOLOGICAL PERSPECTIVES ON 2011 TYPHOONS NESAT AND NALGAE

Ty Matejowsky

ABSTRACT

Purpose – *This paper addresses how local retailers remain resilient in negotiating the lead up to and immediate aftermath of two major disasters (Typhoons Nesat and Nalgae) within a developing urban context (Dagupan City, Pangasinan). It highlights the specific mechanisms by which urban traders engage the Philippines' more pervasive and highly resilient "culture of disaster" vis-à-vis conditions of chronic natural hazard.*

Methodology/approach – *This study relies predominately on the traditional anthropological techniques of participant observation and informal/semi-structured interviews to gather relevant project data.*

Climate Change, Culture, and Economics: Anthropological Investigations
Research in Economic Anthropology, Volume 35, 239–262

ISSN: 0190-1281/doi:10.1108/S0190-128120150000035010

Supplementing these two core methods are findings derived from secondary sources like local and provincial newspapers, government records, public and university libraries, and census findings.

Findings – *Findings suggest that a continual cycle of disaster impact and response does not overtly affect small retailers' entrepreneurial initiative. It becomes clear that a persistent threat of natural hazards fosters a rather fatalistic sense of self-reliance.*

Research limitations/implications – *Study was designed and funded as a quick-response study; therefore, the research timeframe was rather compressed and the informant pool somewhat limited.*

Social implications – *The Philippines is widely recognized as a "culture of disaster" given its volatile position along the Pacific's "Ring of Fire" and "Typhoon Alley." This distinction assumes added dimension as the effects of global climate change become increasingly pervasive at the local level.*

Originality/value – *This paper adds ethnographic detail to a growing body of data on small business resilience within disaster prone areas of the Global South amid intensifying global climate change.*

Keywords: Philippines; climate change; merchants; natural disasters; resilience

INTRODUCTION

Few nations are arguably more closely associated with chronic natural hazard[1] impacts and subsequent disaster recoveries than the Philippines. The country is widely recognized as a "culture of disaster" (Bankoff, 2002) given its volatile position along the Pacific's "Ring of Fire" and "Typhoon Alley." Each year millions of Filipinos are subject to a wide range of severe seismic and climatological activities that frequently disrupt the normative patterns of workaday life. While these catastrophes can generate considerable socioeconomic dislocations that take months, if not years, for affected populations to overcome, local responses to these conditions are not typically viewed as anomalous. If anything, dealing with the consequences of natural hazards has emerged as familiar and quasi-normative aspects of people's lives (Bankoff, 2002, pp. 1–2). Disaster-related disruptions and consequent recoveries are so pervasive, in fact, that they have become embedded into the schema of daily life to the point where locals are well adapted to their occurrence (Sabater, 2008).

With an almost ingrained ability to adapt and overcome setbacks stemming from recurring natural hazards, today's Filipinos occupy a somewhat paradoxical position when it comes to confronting the escalating challenges presented by global climate change. That is, while few populations are as advantageously equipped or, at the very least, amply seasoned in dealing with the socioeconomic ramifications of frequent natural disasters than those living in the Philippines, few populations are also as highly vulnerable to the meteorological effects of appreciable but significant rises in worldwide temperatures as the devastating case of Super-Typhoon Haiyan in November 2013 makes clear. Rising sea-levels and oceanic warming increase the intensity of storm systems which, in turn, cause more damage to exposed population centers and human structures.

Unlike other parts of the world where exposure to these extreme climatological conditions is largely new, the Philippines has a long history of contending with such severe geophysical events and their residual effects. With the world's fourth longest coastline and nearly 20 typhoons striking the archipelago each year, the country and its inhabitants stand poised for even more dislocation as future storm surges emanating from these oceanic disturbances are predicted to strengthen amid intensifying global climate change (Dasgupta et al., 2009). Significantly, it is no longer just local residents situated within exposed coastal zones and low-lying areas that are especially vulnerable to these unfolding developments as typhoon surges now move further inland over larger swaths of populated regions.

As with other parts of the tropics, these adverse climatic conditions and corresponding responses tend to gain maximal expression during the country's rainy season. This prolonged period of severe weather activity which roughly stretches from June to December is characterized by a cycle of repeated tropical cyclones of varying strength and destructive power. Given the increasingly palpable effects of global climate change over recent years, no region within the archipelago is now spared from the devastation wrought by these frequently massive storm systems. For example, the main Philippine island of Luzon endured the destructive force of two Category Three typhoons that hit the region within days of each other in late September and early October 2011. Typhoons Nesat and Nalgae, known locally as Pedring and Quiel, respectively, battered towns and cities across Luzon with successive waves of high winds, storm surges, torrential rains, and catastrophic flooding. Especially hard hit was the country's third most populous province, Pangasinan, which lies in the north-central part of the island and suffered over ₱700 million ($1 USD = ₱54 thus, approximately $16 million USD) in initial typhoon damages (Manantan-Yparraguirre, 2011b).

As the province's largest coastal urban center, Dagupan City bore much of the brunt of Nesat and Nalgae's devastation as its geographical location along a stretch of low-lying land where three major rivers converge and empty into the Lingayen Gulf offers very little in the way of natural protection. Dagupan is located some 210 km north of Metro-Manila and stands as one of north-central Luzon's foremost commercial, educational, and financial centers, with some 130,000 residents living in approximately 25,000 households spread across 31 precincts. For local residents, Typhoons Nesat and Nalgae emerged as just the latest in a series of major catastrophes to strike the city center and surrounding areas with seeming regularity.

Over the past three decades, Dagupan experienced its fair share of major seismological and meteorological disasters including multiple Category Three and Four typhoons as well as a major 7.8 earthquake in July 1990. Among other things, these geophysical events significantly disrupted urban economic activities; even to the point where some shopkeepers and vendors seriously considered abandoning downtown for more secure sites elsewhere (Matejowsky, 2002). The local trade community is able to surmount many of these disaster-related challenges with an ostensible tenacity that has proved successful for both retailers and the city as a whole (Matejowsky, 2009). Since the mid-1990s, greater Dagupan has experienced sustained levels of post-earthquake commercial development including the opening of three large-scale shopping malls and several multinational fast-food restaurants in and around downtown (Dannhaeuser, 2004). This period of formal sector economic growth has not occurred without setbacks as efforts to develop parts of the city are punctuated by bouts of widespread street flooding and other ill-effects stemming from repeated typhoon activity and inadequate public infrastructure. Without upgrades to local storm drainage and sewer systems, these kinds of impediments to urban economic growth will likely persist.

As in other provincial Philippine cities, local commerce in Dagupan generally conforms to the dichotomous formal/informal economic model whereby a limited but growing number of medium-scale formal enterprises operate alongside scores of small-scale informal ventures. Dagupan's formal sector of urban retailing is pretty much dominated by a small but influential community of Chinese-Filipinos whose local roots date back to the

19th century. While comprising less than 5 percent of Dagupeños, this tight-knit group has remained at the forefront of downtown's street store trade for generations – especially in sub-sectors like groceries, textiles, hardware, and appliances – and on average their firms are more highly capitalized than those operated by their ethnic Filipino counterparts. In part, this business supremacy is facilitated by the coordinated efforts of the local Chinese Chamber of Commerce and the more recent participation of Chinese-Filipino business owners in city government and shopping mall development (Dannhaeuser, 2004).

Everyday consumption and employment needs for most Dagupeños are variously met through the informal trade sector. Although this informal economic sphere has wider participation by native Filipino market vendors and mom-and-pop shopkeepers, it also includes numerous ethnic Chinese street store operators. Smaller street store enterprises maintain sales staffs numbering around five to ten employees, while larger ventures such as several city supermarkets and appliance stores employ two dozen or more. Although both trade sectors were adversely affected by the initial and lingering effects of Typhoons Nesat and Nalgae in fall 2011, those retailers operating less formal ventures were arguably more exposed to the storms' devastation thanks to their specific levels of capitalization and geographical concentration in the city's flood-prone central business district.

Indeed, the powerful storms and their immediate aftermath brought downtown commerce to a virtual standstill as local vendors and street store merchants scrambled to protect their operations from Typhoons Nesat and Nalgae's more calamitous effects. The pervasive floodwaters that remained largely undiminished in the days after Nalgae moved westward into the South China Sea proved especially troublesome for Dagupan's established retail community over subsequent weeks. These conditions were exacerbated by the fact that most downtown businesses were never fully restored to their prior uniform elevation following the rapid liquefaction of ground soil and geological sediment triggered by the aforementioned 1990 earthquake (Matejowsky, 2009). Since physical repairs were only partially implemented over intervening decades, the city's street-level urban fabric lacks consistent structural uniformity with numerous sunken sales floors and only minimal safeguards against chronic flooding. These building irregularities become particularly acute during the rainy season and even more so following a major typhoon when many parts of downtown can remain submerged for days or weeks on end.

RESEARCH OBJECTIVES

Against this backdrop of ongoing disaster vulnerability, undercapitalized infrastructure, and intensifying global climate change, questions arise about how affected merchants negotiate the logistical complexities and economic disruptions brought on by such severe though not uncommon meteorological events such as Typhoons Nesat and Nalgae. Significantly, how do small- and medium-scale enterprises (SMEs)[2] in developing contexts affected by climate change such as the urban Philippines remain resilient and commercially viable amid the persistent threat and impact of these and other natural hazards (Dyer, 1999, pp. 282–283; Mula, 1999; Resurreccion-Sayo, 1991)? To what degree, if any, do they rely on external sources of assistance such as private insurance or government disaster relief to insulate their economic vulnerabilities and expedite their recoveries? To better understand how the looming prospect and impact of natural catastrophes affect local merchants' business decisions and strategies, I consider the recent experience of Dagupan City in the lead up to and aftermath of Typhoons Nesat and Nalgae.

When Dagupan's geographical exposure to various types of natural hazards is considered alongside its status as an important commercial hub for north-central Luzon, it is probably unsurprising that several studies documenting the effects of disasters on local commerce have been published over the years (Dannhaeuser, 1997, 2004; Matejowsky, 2002, 2009). So far, the post-disaster anthropological research conducted on Dagupan's SMEs has taken place invariably years after the fact, and as parts of larger and more comprehensive fieldwork projects. The interviews and other ethnographic techniques employed in these earlier efforts only go so far in detailing city merchants' experiences in confronting such extreme geophysical events. Moreover, studies within this body of work have never directly examined the impact of a major typhoon, much less two successive storms, amid intensifying global climate change on local retail life.

The approaches that urban SMEs employ to retain their viability under such extreme and persistent conditions have, thus far, not been fully explored ethnographically. Anthropological consideration of these strategies stands to contribute to the social scientific literature on natural disasters and climate hazards in various ways. At a more general level, it elucidates how SMEs negotiate a major disaster's immediate and secondary effects within a developing urban context. More specific to the Philippines, it deepens understanding about the specific mechanisms by which SMEs engage the country's highly pervasive and resilient "culture of disaster" vis-à-vis conditions of chronic natural hazard. Finally, and perhaps most importantly, it helps identify opportunities for public/private collaboration that can link local business networks to external institutional resources in an effort to expedite natural hazard recoveries. Indeed, the value derived from developing partnerships of this type extends beyond the Philippines to other parts of Southeast Asia and the Global South where the effects of global climate change become disproportionately felt.

Merchant coping strategies within the context of chronic natural disasters are particularly intriguing given the importance of SME operations to the Philippine economy. SMEs comprise 99 percent of all business establishments in the Philippines and 60 percent of the country's exporting firms (Department of Trade & Industry, 2009). Similarly, they employ nearly 55 percent of the Philippines' labor force and supply 30 percent to the total domestic volume sales (Department of Trade & Industry, 2009). Such figures arguably assume a deeper significance given the fact that SMEs mainly operate under conditions of undercapitalized public infrastructure, insufficient institutional support for disaster preparedness and recovery, and an increasingly neoliberal or privatized political economy whereby government oversight and expenditures on public works are steadily eroding (Balicsan & Hill, 2003). In this way, government economic policy or the lack thereof can play a significant role as to how and in what way SME disaster recoveries ultimately unfold.

OVERVIEW AND ISSUES

After providing a brief description of the data elicitation strategies utilized in this study, I outline some of the key characteristics that distinguish those SMEs within Dagupan's retail trade community profiled over subsequent sections. Next, I consider their respective experiences related to natural hazard preparation and mitigation, in general, and the particular challenges they faced grappling with the effects of Typhoons Nesat and Nalgae, more specifically. I then synthesize these findings into a more comprehensive account of local merchant coping strategies so that the individual experiences of examined SMEs can be critically evaluated against one another. Finally, I summarize my study results and discuss potential future research trajectories as well as some tentative policy recommendations that have implications both for the urban Philippines and communities across the Global South as the immediate and lingering effects of global climate change become more and more widespread.

Geophysical events of the magnitude of the 1990 earthquake or Typhoons Nesat and Nalgae are often difficult to anticipate with any real sense of accuracy, much less actively investigate with a long-gestating plan of study. While the extreme, albeit familiar, conditions of natural hazard preparation and recovery offer a viable opportunity to capture important and timely insights about the mitigation practices that urban SMEs have developed over time, they are not always conducive to the holistic and comprehensive research designs that cultural anthropologists tend to favor. Bearing this in mind, a few caveats about the utility and scope of this research effort are probably warranted as it was designed and funded as a quick-response study[3] whereby intensive fieldwork is conducted among affected populations within a compressed timeframe shortly after a disaster strikes, so as to effectively capture perishable data before they disappear. Such qualifications are useful as they not only help delineate the workable boundaries of this ethnographic approach but also establish some reasonable expectations about the kinds of findings that are achievable within the specialized circumstances of post-disaster contexts.

First, the need to document Nesat and Nalgae's impact on Dagupan's retail trade community proved rather time sensitive in the sense that this back-to-back occurrence provides a novel opportunity for anthropological research on urban commerce within the context of recurring natural disasters. Second, data collected in the weeks following a major catastrophe are arguably more highly valued as affected retailers are presumably less emotionally distraught from the disaster's initial impact and are already some way into the recovery process. As such, they are likely more candid and detailed in answering ethnographic interview questions about the strategies they employ when contending with disaster preparedness and mitigation than they would be over subsequent months and years. Third, it is the spontaneous and chaotic nature of catastrophes like Typhoons Nesat and Nalgae that largely determines the timing and feasibility of this and other quick-response studies. Since these disasters can strike with little or no prior warning, assessing their local impact in a timely and effective manner can prove quite challenging. When considered all together, these caveats help define the research parameters of merchant resiliency and local conditions within a pervasive context of recurring natural hazard in Dagupan City.

METHODOLOGY AND INFORMANT PROFILE

Firsthand information about Nesat and Nalgae's impact on Dagupan's retail trade community and local merchants' responses to these powerful typhoons was collected over several weeks in December 2011. This study relies predominately on the traditional anthropological techniques of participant observation and informal/semi-structured interviews to gather relevant project data. Supplementing these two core methods are findings derived from secondary sources like local and provincial newspapers, government records, public and university libraries, and census findings.

With regard to participant observation, I surveyed and visually documented the typhoons' damage to local infrastructure and commercial life. Data collected through this technique were enhanced as the physical impact of both disasters was still readily evident at the community level. With regard to interviewing, I talked with nine SME owner/operators and urban vendors as well as several other local informants in sessions lasting approximately 30 minutes to an hour and half. The timing of these discussions proved fortuitous as enough weeks had passed since Nestat and Nalgae that events still remained fresh in their memories with the various stresses of storm preparation and recovery by now giving way to a more considered reflections of recent happenings.

Interviewed traders represent a wide array of local retail sectors including those dealing in textiles, school supplies, baked goods, flowers, groceries, and food service. While all of the interviewed merchants' retail ventures are situated in and around downtown Dagupan, there remains modest to significant variation in their respective capitalizations. For example, the most highly capitalized retail venture documented in this study is the local bookstore dealing in school supplies and sporting goods, while the least capitalized of these examined commercial ventures are those maintained by small food service operators. In terms of ethnicity, two of the nine merchants interviewed are fourth or fifth generation Chinese-Filipinos, with remaining seven enterprises owned and operated by native Filipinos (see Table 1).

Even though this pool of informants is by no means exhaustive, selected interviewees do represent a viable cross-section of Dagupan's downtown trade community from which important findings can be derived. Accordingly, the questions posed to these merchants not only deliniate important demographic and operational characteristics, they also help tease out details about the prevention and mitigation strategies employed in the time bracketing Typhoons Nesat and Nalgae (see the appendix for interview questions).

Table 1. Informant Characteristics.

Name	Type of Business	Ethnicity	Classification	Number of Employees	Years at Current Location
Benson Marketing	Textile	Chinese-Filipino	Downtown street store	10	42 years
Kwong Tay Bakery	Bakery	Chinese-Filipino	Downtown street store	n/a	Since before World War II
Marigold Bookstore	Bookstore	Filipino	Downtown street store	27	n/a
E.C. Daodana's Rice Store	Rice store	Filipino	Market stall	2	Since mid-1960s
J.P.'s Restabar	Restaurant/bar	Filipino	Market stall	3	Two years
Lani's Pharmacy	Pharmacy	Filipino	Market stall	2	Three years
Teresa Manaois Little Grocery	Grocery	Filipino	Downtown street store	2	Since late 1980s
Blossoms and Lace	Flower shop	Filipino	Downtown street store	2	15 years
Darwia Bakeshop	Bakeshop	Filipino	Downtown street store	8	3 years

MERCHANT PREPARATIONS FOR TYPHOONS NESAT AND NALGAE

In late September 2011, Dagupeños were greeted with the rather ominous news that a major typhoon was rapidly forming in the Pacific Ocean, several hundred kilometers to the east of the Philippines. Such reports were neither that unexpected nor unusual as the country had already experienced its share of severe meteorological activity related to the cyclical patterns of the annual rainy season. Yet, however unsurprising the prospect of another episode of extreme weather appeared, it did not make the reality of the situation any less daunting for local urbanites. In fact, as the effects of global climate change such as typhoon strength have become more pronounced over recent years, these types of events have assumed new levels of volatility and intensity for affected Filipinos.

Residents originally thought that only the storm's weaker outer bands would skirt the city as the typhoon's predicted forecast through the Luzon Strait was initially oriented more toward Taiwan than the northern Philippines. These hopes were quickly dashed as a high-pressure system bearing down from the north shifted the course of the storm, now formally known as Nesat, into a more direct alignment with Dagupan City and surrounding areas. Compounding this unwelcome turn of events was the news that another comparable sized typhoon, Nalgae, was rapidly developing on the heels of Nesat along an almost identical trajectory. As the back-to-back typhoons approached landfall, many Dagupeños erroneously, but not altogether unreasonably, feared that the two storm systems would merge together into a catastrophic super-typhoon.

Heeding government warnings broadcast hourly on local radio and television, Dagupeños had little less than a day to brace for the fast-moving typhoons. With the clock now ticking, SME owner/operators began taking the necessary steps to protect their business interests from the advancing storm systems. For urban shopkeepers and vendors, these efforts were not really that unprecedented as few lacked experience in dealing with these types of disasters over the years. For downtown merchants, especially those located along the city's flood-prone main business thoroughfare of A.B. Fernandez Avenue, storm preparations effectively meant securing store properties and safeguarding merchandise and equipment from the anticipated deluge of floodwaters and gale force winds that would pound the city over the coming days.

For example, a 69-year old Chinese-Filipino shopkeeper who I interviewed in December 2011 reported moving most of his inventory of textiles to the second floor of his rented building space in the hopes of protecting his stock from severe water damage. Even with the help of family and sales staff, his preventative efforts took almost an entire business day and were only partially completed before being cut short by Typhoon Nesat's imminent arrival. Having been at his present location since 1969, he has endured numerous disasters and their corresponding recoveries. Previous losses in store merchandise compelled him to upgrade his premises over the past decade against such eventualities; even going so far as to outfit his shop with raised metal display shelves that stand nearly a foot off the ground so as to mitigate against inevitable street flooding.

His more highly capitalized neighbor, a Filipina in her early 50s, whose family-owned bookstore employs over 27 workers and offers a wide array of products including school supplies, sporting goods, and greeting cards, took such preventative measures a few steps further than the textile dealer. To protect her store merchandise and facilities from these kinds of natural hazards, she had previously purchased both an electric pump to expedite the removal of floodwaters that sometimes stands ankle-deep and private insurance to cover some of the business losses that invariably arise whenever a major storm system strikes Dagupan City. Of all of the SME owner/operators whom I interviewed, she was doubtless the most willing, if not financially capable, to plan ahead for various future disaster contingencies.

Since so few downtown shopkeepers had the time or resources to effectively protect their sunken sales floors with sandbag barriers, plywood, sheet metal, or other durable materials, the threat of pervasive store flooding became pretty much a foregone conclusion as Typhoon Nesat made landfall. Accordingly, the only real option was to wait out the storms and try to mitigate their most destructive aspects through some of the aforementioned preventative measures. Food and dry-goods vendors located in the city's various public and private marketplaces were perhaps even more susceptible to merchandise losses compared to their local street store counterparts. Such vulnerability was predicated, in part, on their lack of additional or elevated storage space where stall inventories could be temporarily relocated from the imminent arrival of floodwaters.

With only modest capitalizations, the plight of market vendors was further complicated by the fact that the option of purchasing private insurance available to the bookstore owner and other downtown shopkeepers was entirely cost prohibitive for them. As such, coping with the one-two punch of Typhoons Nesat and Nalgae for local petty traders effectively became a matter of improvization. For example, one Filipino rice vendor in his mid-60s who operated a small rented space with his wife in a private marketplace near A.B. Fernandez Avenue was forced to erect an impromptu barrier out of available materials to guard against the looming threat of stall flooding. His attempts to erect an impermeable wall along the front of his market stall front by stacking full 20-kg rice sacks atop one another almost like a makeshift sandbag barrier proved only partially successful.

THE AFTERMATH OF TYPHOONS NESAT AND NALGAE

Dagupan City bore much of the brunt of the dual storm systems as they barreled through northern Luzon over several days in late September and early October, bringing torrential rains and battering winds strong enough to uproot trees and knock out full power service for almost a week. A gap of roughly 12 hours separated the back-to-back typhoons, leaving local residents little time to make significant repairs or adjustments stemming from Nesat as the next oceanic disturbance approached. By the time Typhoon Nalgae had passed over the region, most of the fears related to pervasive street and store flooding had been realized. Just to get to their stores, vendors and shopkeepers along A.B. Fernandez Avenue had to wade through knee deep floodwaters that assumed increased depth whenever high tide rolled in.

Complicating the ability of SME owner/operators to begin cleanup operations was the ill-timed release of excess water from the spillway gates of the nearby San Roque Dam. This reservoir embankment barrier was constructed in eastern Pangasinan along the Agno River and commenced operations in 2003. The Agno River is the Philippines' third largest river and feeds into various downstream tributary systems before emptying into the Lingayen Gulf near Dagupan City. Over the years, the erratic effectiveness of dam operators in handling the urgencies associated with major typhoons and heavy rains has proven continually problematic for city traders (Northern Watch, 2011).

As many of my informants told me it was the decision to discharge excess waters all at once from the San Roque Dam rather than allow gradual preemptive spillway releases, more than the storms themselves, which prolonged the storms' devastation and caused myriad problems for them in terms of water damage and attendant cleanup efforts (Manantan-Yparraguirre, 2011a). Compounding this official miscalculation was the lack of effective or regular river dredging by various government entities. The Dagupan City government had recently purchased a large dredging vessel to clear the heavily silted waterways that traverse the area in and around downtown. By fall 2011, such improvements had only made minimal progress, doing little to ease river obstruction, much less decrease the risk of area flooding triggered by typhoons such as Nesat and Nalgae (Micua, 2011).

This apparent lack of foresight on the part of authorities on how to properly release dam water and channel it safely through populated areas also underscores the ambivalence, if not outright disapproval, that many SME owner/operators feel toward state, provincial, and municipal officials and their responses to the storms' immediate devastation. One downtown shopkeeper stated that the city government's failure to act right after the dual disasters was probably influenced by municipal officials anticipating the disbursement of calamity funds by the national government or other relief agencies. He believes that local politicians pocket or divert some of these resources to their own personal slush-funds. Another goes so far as to claim that this kind of ineffectiveness by government officials creates a dangerous situation that can easily serve as a pretext for "the declaration of martial law" to restore civil order and prevent looting amid disaster recoveries.

While provincial and local governments did not remain idle in the lead up to and immediate aftermath of Typhoons Nesat and Nalgae, most of their typhoon-related aid efforts were directed toward either rescuing stranded homeowners or providing disaster relief to city households and neighborhoods. As such, local SME owner/operators were more or less left on their own when it came to preparing for recovering from the twin storm systems. While many affected households obtained assistance in the form of food and construction materials from the city's Department of Social Welfare and Development, downtown merchants received no such comparable support or relief goods. This kind of prioritizing effectively reinforces notions that government office holders essentially look after their own political interests and their wider constituencies and not those of everyday SME owner/operators.

Despite this lack of official support in the days and weeks following Typhoons Nesat and Nalgae, local merchants were able to turn to each other for some degree of assistance, especially in matters related to store security. Neighboring shopkeepers coordinated among themselves to look out for each other's retail properties as the immediate effects of the two storm systems still persisted. A very real fear existed among affected SME owner/operators that local criminal elements or opportunists would take advantage of the chaos and security lapses prevalent during the post-disaster context to loot their respective businesses of valuable goods and equipment. Given the significant flooding and power outages that brought downtown commerce to an effective halt, such concerns were entirely reasonable. Accordingly, most independent traders along A.B. Fernandez Avenue placed one or two employees on lookout once the worst of the storms had passed and preliminary storm recoveries commenced. Acting as visible deterrents, if these workers noticed anything suspicious along nearby storefronts they would either take action or report it to the police.

MERCHANT LOSSES AND COPING MECHANISMS

While not as devastating as other storm systems that have hit the area over recent years, Typhoons Nesat and Nalgae still caused considerable dislocations for city vendors and shopkeepers in their collective wakes (see Table 2). Physical damages mainly entailed losses in merchandise and supplies as well as on-site water damage to store fixtures and equipment. Only two informants, a local pharmacist and an independent bakeshop owner, reported sustaining no building damage or product losses resulting from the extreme weather disturbances (see Table 2). Notably, both businesses were housed in properties without sunken sales floors resulting from the 1990 earthquake.

The rest of the SME owner/operators that I interviewed endured slight to significant property damage because of intense street flooding and subsequent water damage. For example, a local florist and gift shop owner had to discard a water-saturated sofa and unsalvageable refrigerator out of health and safety concerns related to mold growth, while shelves and other shop fixtures at the local bookstore had to be thrown away due to severe warping. Most other affected merchants had to deal with ankle to knee-deep standing water, silt runoff, and other persistent aftereffects that lasted for days and even weeks amid chronic power shortages. Since few of the less capitalized merchants had access to motorized pumps, water removal had to be accomplished through more labor intensive methods such as mopping and squeegeeing sales floors.

Beyond damage to store properties and spotty electrical service, nearly all of the traders examined experienced some level of damaged inventory. Some SME owner/operators were forced to throw out goods that they

Table 2. Effects of Typhoons Nesat and Nalgae 2011.

Name	Reported Damage	Days Closed	Time for Normal Business to Return	Reported Financial Losses	Help from Suppliers/ Creditors
Benson Marketing	Store flooded; damaged merchandise	Three days	n/a	₱6,000	None
Kwong Tay Bakery	Store flooded; damaged equipment and bakery ingredients	Three days	One week	₱100,000	None
Marigold Bookstore	Store flooded; damaged merchandise	Two days	One week	₱90,000	Longer payment period; exchanged damaged merchandise
E.C. Daodana's Rice Store	Stall flooded; damaged merchandise	Three days	One week	₱40,000	None
J.P.'s Restabar	Stall flooded	Three days	Two months	₱15,000	None
Lani's Pharmacy	No reported damage	Two days	One week	₱20,000	None
Teresa Manaois Little Grocery	Stall flooded; damaged merchandise	Three days	One week	₱9,000	Allowed placement of emergency orders
Blossoms and Lace	Store flooded; damaged merchandise and equipment	Three days	One week	₱5,000	None
Darwia Bakeshop	None	Two days	One week	₱15,000	Delayed credit payments for 15–30 days

deemed no longer vendible, especially those selling food and other perishable items, while others found ways to salvage some merchandise for potential resale to customers. For instance, the Chinese-Filipino textile dealer simply air-dried and respooled all of his waterlogged fabrics at home before restocking them on store shelves several weeks later. Somewhat less dubiously than her neighbor, the female bookstore owner was able to financially recoup some of the losses she experienced with damaged store stock as she had previously purchased flood insurance for such contingencies.

Perhaps most extreme in terms of material losses is what occurred to a 69-year old Chinese-Filipino bakery owner whose family business has been operating in downtown Dagupan since before World War II. With a sunken sales floor and heavy equipment difficult to move on such short notice, the store was particularly vulnerable to street flooding. Not only were large amounts of sugar and flour destroyed as a result of Typhoons Nesat and Nalgae, but some expensive baking equipment was ruined; thereby making an immediate financial recovery for him all the more difficult. For over a week, he expressed feeling powerless to do anything to ameliorate the situation, stating that "all you could do was eat and sleep." Ultimately, he reported enduring some ₱100,000 in financial losses as a result of the twin storm systems (see Table 2). Like most other downtown businesses, the bakery was not covered by private insurance so he had to absorb the economic costs himself.

Such property and merchandise losses were compounded by the fact that little or no business revenues were generated during the time bookending Typhoons Nesat and Nalgae. Rather than selling their respective wares to city shoppers, most of the SME owner/operators I examined reported having to close their store/stall operations for two to three days in early October as the back-to-back storm systems slammed the region. Severely cutting into their monthly net profits, these losses were further exacerbated by rather tepid consumer spending that followed over subsequent days and weeks.

Understandably, local consumers were either too preoccupied with their own post-disaster recoveries to patronize downtown retail establishments or unable to spend money on anything not directly related to their immediate recovery needs. Thus, it took at least a week before the examined SME owner/operators began to notice a normalizing of local business activities. Considering that these kinds of interruptions are not isolated events but, rather, occur each year with near regular frequency in Dagupan City and elsewhere across the urban Philippines, it is easy to see how annual sales never reach their optimal potential amid such severe external weather impacts.

The willingness of some product suppliers and creditors to negotiate or modify established business relationships with their retail clients somewhat assuaged the complications and setbacks suffered by Dagupan's merchant community in the wakes of Nesat and Nalgae. At least three of the nine SME owner/operators whom I interviewed were able to secure workable arrangements from those with whom they purchase stock to get through the disruptions brought on by the dual storm systems (see Table 2). Mainly, this entailed extending the time available to repay credit obligations or exchanging waterlogged merchandise for undamaged stock. Apparently, maintaining smooth working relationships with local retailers in times of difficulty is something suppliers view as mutually beneficial over the long term.

DISCUSSION

Documenting Dagupan's merchant community in the wake of Typhoons Nesat and Nalgae offers some revealing insights into how SME owner/operators negotiate the complexities of disaster preparation and recovery amid escalating global climate change. Collected data for this research project reveal several notable, and in some cases unexpected, results. Tentative findings suggest that a continual cycle of disaster impact and response does not overtly affect small retailers' entrepreneurial initiative. Based on research findings in the weeks subsequent to the dual storm systems, it becomes clear that a persistent threat of natural hazards fosters a rather fatalistic sense of self-reliance. That is, most see this kind of commercial environment as an integral aspect of earning a livelihood in today's urban Philippines. Such sentiments are prevalent even though most of these SME owner/operators do not really identify themselves as risk-takers or entrepreneurs. It seems that Dagupeño merchants take the disaster-related displacements they periodically encounter in stride; viewing these setbacks as part of a normal course of events. Whether or not these attitudes will persist remain something of an open question as the meteorological and oceanic disturbances impacting the Philippines increase in intensity as the global climate change continues to escalate.

Moreover, unlike the findings of other researchers in the Philippines (Bankoff, 2007), local SME owner/operators do not really rely on external sources of public or private assistance such as government disaster relief or flood insurance to insulate their economic vulnerabilities and expedite their recoveries. While affected merchants seemingly do not expect much in the way of outside assistance in post-disaster recoveries from local, provincial, or national government organizations, they see this more as another indication of policymakers' inability to meet the basic needs of everyday Filipinos. If anything, the overriding sentiment with regard to government aid in the wake of Typhoons Nesat and Nalgae was one of ambivalence.

Many noted that the government did an adequate job of providing relief to poor households across Pangasinan and surrounding areas, but did little or nothing to help local businesses get back on their feet. Complicating matters was the persistent belief among affected merchants that government officials actually exacerbated recoveries by recklessly releasing overflow floodwater from the nearby San Roque Dam. Such pervasive attitudes among urban SME owner/operators about government ineffectiveness potentially complicate future disaster recoveries as the Philippine government begins to prioritize the impact of global climate change on towns and cities across the archipelago in the wake of November 2013 devastating Super-Typhoon Haiyan.

Given the overall lack of external assistance for affected merchants, insights about the risk aversion/management strategies and coping mechanisms employed by Dagupeño traders in the immediate wake of Typhoons Nesat and Nalgae did not emerge the way I initially anticipated. In particular, direct support to retailers from government agencies or other institutions did not appear to play a major role in the post-disaster recoveries of most involved SME owner/operators. Only the aforementioned bookstore owner had formal insurance even though this type of financial coverage is available for purchase by all downtown merchants. Therefore, an anticipated social network analysis of formal support linkages that I planned to implement among local retailers did not really prove feasible.

Despite this unanticipated outcome, ethnographic information was collected about how local merchants prepared for the oncoming storms and recovered from their devastation. In response to the advancing approach of Typhoons Nesat and Nalgae, all of those interviewed took precautions to protect their business operations from the storms more devastating effects. These safeguards mainly entailed protecting store equipment and merchandise from the impending arrival of dangerous floodwaters. Given downtown Dagupan's low lying position not far from the Lingayen Gulf and the rapidly approaching storms, it became essential to transfer such items to higher ground before they became waterlogged.

For example, the ethnic Chinese merchant who deals in textiles not only moved much of his store stock upstairs, he also utilized previously installed raised metal shelving to keep merchandise as dry as possible amid the imminent prospect of rising floodwaters. Similarly, the bookstore owner purchased private insurance and an industrial water pump in preparation for persistent disaster impacts. After the storms passed, local SME owner/operators began a recovery process that took days, if not weeks, to complete. Mainly, this entailed getting their respective businesses back to full operational capacity. Standing water removal and the disposal or repair of damaged merchandise, supplies, and equipment was of particular importance for local trader in the immediate aftermath of Typhoons Nesat and Nalgae.

Interviewed SME owner/operators also provided ethnographic information about property losses, financial setbacks, and the respective roles of suppliers and customers in the post-disaster context. Findings related to these issues offer some possible research trajectories for future ethnographic disaster research in Dagupan City and elsewhere across the urban Philippines. First, affected merchants lost anywhere from one week to two months of normal income from the two typhoons. While this may appear inconsequential, one week represents a 2-percent loss in annual sales, and some experienced greater losses while others lost less. Since urban communities like Dagupan City are subject to multiple natural hazards each year, it stands to reason that local commerce is continually underperforming due to the displacements brought on by these major geophysical events. Quantifying such losses over a larger research sample would be one approach to better understand how local retailers grapple with immediate and secondary effects of major natural hazards.

Second, one-third of interviewed merchants were able to craft working arrangements with their suppliers to offset the losses brought on by Typhoons Nesat and Nalgae. The fact that many SME owner/operators did not report similar agreements with those from whom they purchase stock raises questions about how suppliers can play a more effective role in disaster recoveries. That is, it may be worth investigating further how the quality of supplier–client relationships can be strengthened to mitigate the financial hardships experienced by affected merchants in the days and weeks after a disaster strikes.

Third, the decline in customers that resulted in loss of a couple of weeks' income was largely due to the complexities of disaster recovery in the wake of two major typhoons. Again, future research endeavors in Dagupan that explore the variation in the return of customers, and the extent to which customer relations played a role in return of customers and variation in overall losses is worth considering.

CONCLUSION

My prior fieldwork in Dagupan dating back to the late 1990s suggests that downtown shopkeepers and vendors exhibit considerable resourcefulness when confronting the immediate and secondary effects of major natural disasters based on their varied utilization of the Philippines' pervasive "culture of disaster" (Matejowsky, 2009). That said, I have been unable to observe them deal with a major disaster, much less two, so soon after the event. Being there only some seven weeks after Typhoons Nesat and Nalgae, provides a more immediate context for examining how SME owner/operators contend with the complexities of disaster recovery. Specifically, the risk aversion/management strategies and coping mechanisms employed by Dagupeño traders in the time bracketing Typhoons Nesat and Nalgae offer a valuable opportunity to understand how local capital is locally cultivated and engaged, if at all, within post-disaster recoveries. Such abilities become especially important as the effects of global climate change serve to increase the intensity and volatility of future oceanic disturbances to which coastal communities like Dagupan City are vulnerable.

In many ways, this fieldwork effort examining the risk aversion/management strategies and coping mechanisms employed by Dagupeño traders following Typhoons Nesat and Nalgae largely reaffirms my past research findings on the ways that merchants operate more generally within the Philippines' "culture of disaster." Beyond the Philippines, my research findings may also provide insights about policy opportunities in the disaster prone areas of the Global South. Techniques used by small merchants without access to formal institutional assistance may be incorporated into existing policies in other parts of what was once called the Third World, thus, better accounting for relevant needs and capacities at the grassroots level. Despite support provided by international relief agencies like the International Red Cross, relatively little is known about the extent to which or how affected SME owner/operators prepare for and mitigate the more destabilizing effects of major catastrophes (Davidson, 2002). Arguably, even less is known about how SMEs in post-disaster contexts employ less formal forms of support, such as unrecorded loans or credit, unpaid kin-based labor, financial gifts, use of equipment or workspace, shared access to markets, and pooling of resources to reestablish themselves following a disaster. Determining what effective combinations of formal and informal support are sought out and/or accepted by SME owner/operators in a post-disaster setting, particularly independent urban merchants, will likely go a long way toward addressing the knowledge gap that currently persists regarding the variation in resourcefulness and resilience of SMEs in the Global South and elsewhere.

NOTES

1. As a matter of clarification, I use the terms "natural disaster," "natural hazard," and "natural catastrophe" interchangeably.
2. In this work, SMEs are classified in respect to the definition utilized by the Philippine's Department of Trade and Industry (Department of Trade & Industry, 2009). Such a classification may unintentionally gloss over differences in geographical location and type of industry.
3. This research effort was supported through a Quick Response Grant from the Natural Hazard Center at the University of Colorado at Boulder which provides funds for researchers to quickly travel to disaster-affected areas to capture perishable data.

ACKNOWLEDGMENTS

The author wishes to acknowledge the insightful contributions of Josie Gonzalez, Mario Granada, Martha Heine, and Lorena Parker Matejowsky to this work. A particular thanks goes to Eric C. Jones for his valued feedback and involvement with the design and analysis of this research endeavor. Moreover, gratitude is extended to the University of Central Florida (UCF)'s Department of Anthropology and the UCF Southern Region for their assistance during this fieldwork project. Finally, deep felt appreciation is offered to the Natural Hazard Center at the University of Colorado at Boulder for their support of this ethnographic undertaking.

REFERENCES

Balicsan, A., & Hill, H. (2003). *The Philippine economy. Development, policies, and challenges.* Oxford: Oxford University Press.

Bankoff, G. (2002). *Cultures of disaster: Society and natural hazard in the Philippines.* New York, NY: Routledge.

Bankoff, G. (2007). The dangers to going it alone. Social capital and the origins of community resilience in the Philippines. *Continuity and Change, 22*(02), 327–355.

Dannhaeuser, N. (1997). The malling of Dagupan. *Philippine Studies, 45*, 151–173.

Dannhaeuser, N. (2004). *Chinese traders in a Philippine town. From daily competition to urban transformation.* Quezon City, Philippines: Ateneo de Manila University Press.

Dasgupta, S., Laplante, B., Meisner, C., Wheeler, D., & Yan, J. (2009). The impact of sea level rise on developing countries. A comparative analysis. *Climatic Change, 93*(3–4), 379–388.

Davidson, O. (2002). Future disaster research: A practitioner's viewpoint on public-private partnerships. In R. A. Stallings (Ed.), *Methods of disaster research* (pp. 377–386). Bloomington, IN: Xlibris.

Department of Trade and Industry. (2009). *Micro, small, and medium enterprises (MSMES).* Retrieved from http://www.dti.gov.ph/dti/index.php?p=321

Dyer, C. (1999). The phoenix effect in post-disaster recovery: An analysis of the economic development administration's culture of response after Hurricane Andrew. In A. Oliver-Smith & S. M. Hoffman (Eds.), *The angry earth: Disaster in anthropological perspective* (pp. 278–300). New York, NY: Routledge.

Manantan-Yparraguirre, L. (2011a). *San Roque Dam releases water to avert spillage. Sun-star pangasinan*, September 30. Retrieved from http://www.sunstar.com.ph/pangasinan/local-news/2011/09/30/san-roque-dam-releases-water-avert-spillage-182224.

Manantan-Yparraguirre, L. (2011b). Pangasinan under State of Calamity. Sun-Star Pangasinan, October 3. Retrieved from http://www.sunstar.com.ph/pangasinan/local-news/2011/07/03/pangasinan-under-state-calamity-164749

Matejowsky, T. (2002). Globalization and retail development in the post-disaster context: A comparison of two Philippine communities. In N. Dannhaeuser & C. Werner (Eds.), *Research in economic anthropology: Social dimensions in the economic process* (Vol. 21, pp. 311–341). Amsterdam, Holland: Elsevier-JAI Press.

Matejowsky, Ty. (2009). When the lights go out: Understanding merchant "Brownout" behavior in the provincial Philippines. In E. C. Jones & A. D. Murphy (Eds.), *The political economy of hazards and disasters. Society for economic anthropology monograph series* (Vol. 27, pp. 291–310). Walnut Creek, CA: Altamira Press.

Micua, L. V. (2011). DPWH Exec Underscores Need to Dredge Pantal River to Solve Dagupan's Flood Woes. Bayanihan, 26 October. Retrieved from http://bayanihan.org/2011/10/26/dpwh-exec-underscores-need-to-dredge-pantal-river-to-solve-dagupans-flood-woes/

Mula, R. P. (1999). *Coping with mother nature. Household's livelihood security and coping strategies in a situation of continuing disaster in tarlac, Philippines.* Wageningen: Proefschrift ter Verkrijging van de Graad van Doctor.

Northern Watch. (2011). *Years after the San Roque Dam Flood, where is the justice?* Retrieved from http://northwatch.wordpress.com/2011/06/04/editorial-41/

Resurreccion-Sayo, B.. (1991). Coping with disaster. *Manila chronicle*, 22 June, p. 10.

Sabater, M. R. (2008). In Ring of Fire, Typhoon Belt: Surviving Disasters, 8 June: 6.

APPENDIX

Interview Questions

1. Name
2. POB
3. Year of birth
4. Education
5. Hometown
6. How long have you held current position?
7. How did you come into this position?
8. What precautions did you take in preparation for Typhoons Nesat & Nalgae?
9. Did you receive notification from the city/provincial/national government?
10. What was the worst aftereffect of the typhoons?
11. Did you suffer any damage to your store? If so, what?
12. Did you suffer any damage to your merchandise? If so what?
13. How did the disasters disrupt your relations with suppliers?
14. Do you have insurance?
15. What kinds of financial losses have you suffered?
16. Have you noticed a decline in customers?
17. When did or do you think business will pick up?
18. What do you think of the city government's response to the disasters?
19. What do you think of the provincial government's response to the disasters?
20. What do you think of the national government's response to the disasters?
21. What kinds of infrastructural improvements need to take place to reduce flooding?
22. On whom do you rely for economic/material support to keep your business operating?
23. What kinds of help did you receive?
24. Why do you think the Philippines endures so many disasters?
25. Do you see natural disasters as just part of doing business in the Philippines?

ENVIRONMENTAL DESTRUCTION, GUARANÍ REFUGEES, AND INDIGENOUS IDENTITY IN URBAN PARAGUAY

Richard K. Reed

ABSTRACT

Findings – *The forests of eastern Paraguayan have been cleared, forcing the indigenous Mbyá-Guaraní to take refuge in cities. Rather than assimilate into the city's underclass as other indigenous people do, Mbyá remain on the margins of the national society and protest their land loss in increasingly public demonstrations against the government. This research points to the historical struggle that the Mbyá-Guaraní of eastern Paraguay have waged against the state to explain Mbyá identity and action in the urban environment.*

Research limitations/implications – *Recent work with indigenous refugees shows that dislocation entails not only a disruption of social ties, but efforts to reestablish identities and relations in their new conditions. This research explores the interplay of urban conditions and historic struggles in the development of these new indigenous identities for the Mbyá-Guarani.*

Climate Change, Culture, and Economics: Anthropological Investigations
Research in Economic Anthropology, Volume 35, 263–292

ISSN: 0190-1281/doi:10.1108/S0190-128120150000035011

Practical implications – *Indigenous refugees in extreme poverty are a growing problem in urban Latin America. Once residents of the forest, these groups squat in vacant lots and scavenge for a living, ravaged by disease, drugs, alcohol, and sex work. This work seeks to identify the factors that lead some indigenous people to integrate into urban society, while others assert themselves against that system.*

Keywords: Environmental refugees; indigenous identity; urban migration; Paraguay

Drivers on the streets of Asunción, Paraguay, were disturbed in the late 1990s when indigenous Mbyá-Guaraní began showing up on corners to beg. Young mothers with sad eyes and babies on their hips walked among the halted vehicles imploring the drivers with outstretched hands. Children in rags and dirty faces scampered between the lanes, impishly tapping on the windows of the surprised drivers. Under the dictatorship of Alfredo Stroessner, Asunción had been as a city of considerable decorum. Poverty was kept out of sight and middle-class norms of propriety were enforced with military effectiveness. Indigenous peoples were relegated to a carefully circumscribed corner of the national consciousness: primitives in the forest, living the humble and time-honored lives of farming and foraging. Native peoples on Asunción's streets attracted considerable attention. The daily papers ran pictures of Mbyá families asleep on the sidewalks and television news covered stories of the young men sniffing glue and young women and boys selling sex.

The reason for the Mbyá presence on the streets of Asunción is both obvious and undeniable. Until recently, they occupied the dense, high forests of eastern Paraguay and adjacent border regions of Brazil and Argentina. Then, in an unprecedented wave of ecological destruction, the forests used by the Mbyá were felled for cattle ranches and soybean fields. Between 1980 and 2009, of Paraguay's forests were cut for agriculture, propelling the country to being the fourth largest soybean producer in the world. This was a remarkable feat for a country the size of California (Huang et al., 2009). Mbyá lands were among those hardest hit by the development. Between 2000 and 2010 alone, 15 Mbyá communities disappeared in the area of greatest soy production (DGEEC, 2012) One Mbyá put it succinctly, "We came here because there was nothing left there." "Ndaipori mba'eve upepe," – absolutely nothing.

The migration of forest peoples to urban areas is happening throughout lowland of South America. For decades, displaced indigenous peoples have appeared on the edges of frontier towns as itinerant laborers. As forests are felled and transport improves between isolated sectors and urban centers, these indigenous peoples find their way onto the busy streets of large cities and national capitals.

Urban life appears chaotic and confusing for all rural peoples, but the movement from forest to cities is a profound disrupture for indigenous peoples. Mbyá define themselves and are defined by others (anthropologists and the state) in terms of their relationship with the forest. They call themselves "Ca'aguygua," literally "people of the forest." Anthropology has a strong literature exploring the importance of place as a social construct in creating meaning and defining cultural identity (e.g., Basso, 1996). Others (e.g., Gupta & Ferguson, 1997) extended this discussion of place and identity to include questions of power, embedding the study of landscape and society in a web of political relations. The process of deterritorialization of a social group, then, is more than a spatial process of dislocation, it is a rending of the fabric of identities, social ties, and power relations (DeLeuze & Guatari, 1983). Thus, to be displaced to the cities is more than a geographical dislocation, it is a social and cultural disarticulation.

Recent work in Latin America attends to this process of urbanization among lowland indigenous groups. McSweeney and Jokisch (2007) argue that the movement of native peoples from their traditional lands to the cities is more than a process of deterritorialization, not simply chaotic and confusing, but a reterritorialization of native peoples in the new urban landscape. Indigenous peoples assert new modes of being in the new environment. Gordillo (2011) shows how Guaraní in Argentina struggle to define themselves and their relation with the state. This calls attention to the process by which Mbyá refugees establish themselves in the city. Do they, or how do they, make the city their own? And in doing so, how does that process shape Mbyá identity?

The Mbyá are only one of the indigenous groups finding their way into Asuncion. The 2012 national census counted almost 2,500 native people in Asunción and the surrounding suburbs (DGEEC, 2012), with some estimates suggesting more than 3,500 individuals (Grupo Sunu de Acción Intercultural, 2012), and additional populations in Encarnación and Ciudad de Este (Velázquez, 2013). An urban survey in 2009 identified eight indigenous ethnic groups: two closely related Guaraní groups (Mbyá and

Avá) from eastern Paraguay and a variety non-Guaraní groups from Chaco regions, scattered in various enclaves throughout the city (United Nations, Economic and Social Council, 2010, p. 34).

These indigenous peoples differ greatly in their process of situating themselves in the urban context. All confront problems of urban poverty, disease and social conflicts in the new settings, but some groups fall into a state of "hypermarginality" (Bessire, 2014) on the fringes of the urban social world, often invisible to the state and nation. The attention garnered by the Mbyá is not simply a product of their being indigenous. Mbyá tend to stand out as a disenfranchised minority. Small family groups squat on abandoned parcels, live in cheap rooming houses, and beg on street corners to subsist. Without recourse to an urban network, indigenous, or state, they find themselves relegated to the very lowest rungs of the urban environment.

Other native people often step down from rural buses and meld into the fabric of the city, hardly noticed as they go about their business in town. The Maká were driven from the Chaco to the banks of the river near Asunción and in 1985 established themselves as a cohesive group in an urban neighborhood. More to the point, other Guaraní from eastern Paraguay move much more easily into the city. Avá-Guaraní find shelter and day labor or scavenge for recyclables in the refuse, garnering a daily wage (if not prosperity). In one notable example, Avá have banded together, elected a leader, petitioned for urban land, and won government recognition as a *Comunidad Indígenas*. Despite that they suffer grave injustices, they establish organized relations with city social services and indigenous affairs offices, acquiring state recognition, health care, and schools.

Surveys accomplished among indigenous peoples in Asuncion show a striking disparity between Mbyá and Avá. While most Avá come together at a few well established sites, Mbyá tend to remain dispersed in small family groups in a variety of informal and often illegal settlements. One non-governmental organization found the four established Guaraní communities in Asunción (Cerro Potý, Kenkuket, Tarumandy-mi, and Nueva Esperanza), to be almost exclusively Avá. The Mbyá avoid all but one (Tarumandy-mi), where they have established a small settlement. In addition, there were 11 places where indigenous groups tended to gather, often occupying abandoned buildings, renting houses, and camping in empty lots. Of these six are exclusively Mbyá, two include both Mbyá and Avá families and only two are exclusively Avá (Grupo Sunu de Acción Intercultural, 2012).

How can we understand the process by which some groups establish themselves in the urban environment, while others fail to find stable relations with the state and city institutions? McSweeney and Jokisch (2007, p. 161) point to the ways in which lowland indigenous people's experience of urbanization can be tied to the very process of dislocation and state integration. They suggest that the experience of the migrants in the rural areas shapes the mode of engagement with the state in the old and new landscape alike. The following suggests that the environmental destruction of rural forests is the immediate cause of the flight of indigenous peoples to Asunción, but to understand the Mbyá mode of engagement with the urban system, one must look closely at the history of relations with the state long before they arrive at the city boundaries.

The following explores three aspects of this process. First, it highlights the history by which the state inserted itself into the forest, and the powers by which Guarani resources were destroyed. Second, this work describes indigenous peoples' differing reaction to state encroachment. Unlike Avá-Guaraní, the Mbyá in the forest maintained a historic and steady resistance to the state. Mbyá refused to participate in efforts to facilitate their assimilation into the state and national society, and were forced to take refuge in the city only at last recourse. Third, the following explores the differing experiences of the urban Guaraní, contrasting the plight of Mbyá refugees with Avá-Guaraní who reside in the settlement called "Cerro Potý." Mbyá resistance, this argues, has become a mode of existence in the urban environment, serving as the means to assert a new urban identity in relation to the state.

THE DESTRUCTION OF THE PARANA ATLANTIC FORESTS

Eastern Paraguay has experienced one of the world's most rapid processes of deforestation. As late as 1973, almost three-quarters of the region was covered by forest. Road building financed by USAID in the 1970s allowed colonists and loggers to flood into previously inaccessible forests. By 1980, Paraguayan ranked fourth in world rates of deforestation, with 3.5% of the cover being removed each year (Barnes & Allen, 1985). Rates increased in the subsequent two decades. As shown by Landsat images, the proportion of forested area was reduced to less than half by 1989 and less than a quarter by 2000 (Huang et al., 2009). From 1980 to 1996, the rates of deforestation and the total portion of cover removed were among the highest in the world, with up to 170,000 hectares being cleared annually. A total of 46.7 million hectares of forest was cleared (IPCC, 2007), between 1990 and 2000, an average annual cutting rate of over 4 million hectares that remained constant through 2006[1] (FAO-UN, 2010). Today, it is estimated that only 14% of eastern Paraguay's original forest canopy remains and half of that has been seriously degraded by logging. The remaining forests exist as small and isolated parcels that retain little of the bio-diversity of the original expanses.

To speak in broad terms, three distinct factors destroyed Paraguay's forests. First, cattle ranching transformed vast areas from forest to pasture. The Brazilian economic miracle of the 1960s pushed land sales and clearing across the southern highlands to the Paraguayan border. Brazilian cattle ranchers and their laborers found cheap land and easy access across the *line seca.*

Second, in the 1980s, forest destruction was facilitated by Alfredo Stroessner's weakening grip on the country's power. He sought to mollify the landless masses and counter the Brazilian incursions onto Paraguayan territory by distributing vast forested acreages along the eastern borders to poor mestizo *campesinos* from the densely settled regions near Asunción.

Third, and most profoundly, the world demand for soy created an explosive and land-hungry agro-industry. The soybean was first grown in Paraguay in 1967 and by 1970 54,600 hectares had been planted. This first wave of soy was planted in established wheat fields, and found to rejuvenate the region's weak topsoils. Quickly, investors snatched up the cheap forests along the Brazilian and Argentine borders and, aided by massive road-building projects, converted the region into one of the world's premier soybean exporting zones. While US production stagnated, Brazilian production grew at twice the world rate of increase – and Paraguay's rate of production increase was even higher. By 1987, soybeans covered some 718,800 hectares of Paraguay, more than any other crop, with an annual output of 1 million tons and export revenues of approximately US$150 million. As the world market strengthened, investors flooded in, drying and storage facilities sprung up, the dense forest was pushed into massive windrows, and fields replaced the forest canopy. Today, corporations plant approximately 100,000 hectares of new soy fields each year to reach almost three million hectares planted in 2012 (Markley, 2012).

Mbyá communities were overwhelmed by a wave of environment devastation, deforestation and the erosion, desiccation and water pollution it entails. Many communities suffered a process of serial displacement.

Finding their lands titled to a rancher, the government or corporate agriculture, Mbyá retreated farther into the last vestiges of the forest, and when those were cleared, onto land deemed unsuitable for crops or pasture. In cases where indigenous residents failed to give ground, rural developers quickly resorted to force, invading the vulnerable hamlets with men, guns, and dogs, using bulldozers and fire to destroy houses and fields.

The dislocation inevitably drove the Mbyá to the roads and ultimately small towns. Mbyá became common sights hanging around the edges of rural communities to panhandle or just for the entertainment of watching commercial life. Men worked in the new soybean fields when they could, but there was little labor needed in state-of-the-art mechanized agriculture. Without forest resources, the lack of cash became a critical issue. Adults panhandled on the shoulders of the roads, and children ran onto the hot asphalt calling to cars and trucks. Young people, attracted to the bright lights, found cash in petty crime, cheap drugs, and prostitution.

Finally, as the rural towns themselves proved inhospitable, Mbyá took refuge in one of Paraguay's major urban centers of Paraguay. They caught a ride or bought passage on the buses that ply between Asunción, Encarnación, and Ciudad de Este and descended into the city without food, money, or family. Dressed in threadbare rags, unwashed and unkempt, Mbyá became an extremely visible (and extremely vulnerable) sector of the urban environment. Men lounged in the plazas, young mothers with infants sat disconsolately on the curbs, and young children begged between the rows of traffic. At night they slept on city sidewalks or retreated to vacant lots. In the city, but not of the city, they were strangers in a strange land.

INDIGENOUS PEOPLES AND THE STATE

The destruction of the forest and the displacement of the native peoples of eastern Paraguay have provided them no option but to take refuge in the urban area. But why is it that Guaraní groups have markedly different experiences in the cities? Why do Mbyá generally reject the enclaves other Guaraní have established in Asunción? The answer can be found in the historical relations that the Mbyá have maintained with the national society and international economy. While other Guarani groups, notably the Avá, have negotiated a working relation with the national society through intermittent and at times intense relations with the state, the

Mbyá have refused to be brought more closely into Paraguay's multiethnic society.

From the first years of European colonization in La Plata, settlers and entrepreneurs invaded the Guaraní forests for souls, slaves, and wealth. Jesuits established their famous *reducciones* and sent emissaries throughout the forest to attract converts; Portuguese *bandeirantes* invaded the region from the north east; and Paraguayan mestizos came west from Asunción searching for wealth to be extracted from the forest. The Jesuits and bandeirantes took a serious toll on the population of indigenous people in the forest. By the time that both threats diminished, much of the native population had been killed or assimilated into Paraguayan mestizo society. Others left the region entirely, traveling east with charismatic leaders to the *yvymar'eÿ*, the famous migrations to the "land without evil" chronicled by Nimuendaju (1978), Clastres (1995), and others. Nevertheless, the ancestors of the contemporary Guaraní avoided the missionaries and slavers and ensconced themselves in the deep forests, doing their best to slip unnoticed among the mestizo settlements.

Throughout the 18th and 19th centuries, the relations between native peoples and the larger society remained tense. Commerce in native yerba mate (*Ilex paraguariensis*) attracted mestizo harvesters into the forest. The Spanish historian Juan Francisco Aguirre touring the region at the end of the 18th century reports, "The forest Guaraní alternate between war and peace with the yerba collectors. Killings are few and in general the Guaraní are not bloodthirsty, what they want is to rob [gatherers'] camps for the tools" (Aguirre, 1949, pp. 36–38). The fears, frustrations, and fantasies of the yerbateros about the indigenous residents resulted in violent reprisals against the indigenous communities (Cadogan, 1962, p. 15). By the end of the century, under increasing threat, the large kin-based *tapyí* of forest peoples dispersed into small semi-nomadic bands. Extensive gardens were replaced by small plantings supplemented by gathering and hunting in the vast forests (Cadogan, 1960, p. 136).

The records of the early colonists provide myriad names for the Guaraní-speaking people east of the Paraguay River, but none that correspond to the tripartite division that characterize the population today: the Mbyá, the Avá, and the Paĩ. Most groups were distinguished by local place names (e.g., "Itatines" – people of the Itatĩ River). A second distinction became salient in the 1800s between those who remained ensconced in the forest, the C*a'aguygua*, and those who established relations with the developing Paraguayan nation, the *Guaraní*. As the century progressed, it brought a variety of social and economic pressures that stimulated several ethnic groups among the Guaraní-speaking population. Rather than based in primordial differences within the indigenous population, these identities developed in response to relations with the state, differences in the history of contact with missionaries, colonists, and slave raiders.

In the northern forests of eastern Paraguay, a highly integrated economy of yerba production organized relations between the national society and the Guaraní who came to be known as Avá. These native communities have been in continual and sometimes close relations with the state since the late 19th century. As detailed elsewhere (Reed, 1995), the indigenous communities used regional leaders to enter periodic wage work with the commercial economy, while maintaining (and reinforcing) a sense of their difference from the national society.

Guaraní producers organized teams comprising men from extended kin groups under the direction of indigenous intermediaries. Regional leaders were paid in kind (food, cloth, or tools) for *tercios* (approximately 100 kilograms) of dried yerba. They had the personal ties to keep Guaraní workers at their stations, and might even guarantee the debts of their relatives – making it easier for patrons to collect the product they felt they were due. Over time, these indigenous work leaders were given (or assumed) titles to suggest their authority – *"teniente," "capitas," or "major."* Moreover, mestizo patrons sometimes established liaisons with Avá women and recognized the children from these unions. The progeny, especially the young men, often spent extended periods working with patrons in the yerbales and their homes. As they grew up and returned to live in Avá communities, they wove a complex network of relations between the Avá and the mestizo society of the interior towns.

In the Paraguayan civil war of 1947, Avá-Guarani joined the conflict on both sides, depending on where their personal allegiance to patrons fell. Several of my older friends remember fondly being given rifles and fending off the Colorado advance at the Jejuí River. Others joined the Colorados and carried the fighting all the way to Asunción, where the city was torn apart by unrestrained violence. When the Colorado Party finally took control in 1947, many of the young men who were affiliated with Liberal patrons escaped with their bosses into Brazil. Those who stayed behind threw their loyalties behind whatever caudillo controlled local politics. Many carried a *libretto,* performing unpaid labor fixing roads several times a year instead of paying taxes, and were called upon to vote for the ruling party during elections. They were paid for their allegiance with a modicum of peace and intermittent work.

The history of relations between the state and the Guaraní to the south differed significantly. Yerba was less common in the south-central forests of eastern Paraguay, and the forebears of the Mbyá-Guaraní were more motivated to and capable of defending their forests from mestizo incursions. Even as the large groups had dispersed into the forest, small kin groups maintained ties with a broader kin network known as a "tekoha." While some charismatic leaders had ushered Guaraní in religious migrations to the east, others stayed in the forest and asserted influence and sometimes authority over the dispersed groups in situ. These regional Mbyá leaders became important in mediating relations between the Mbyá and the state. However, rather than promoting long-term relations, Mbyá leaders asserted autonomy. Leon Cadogan, a self-trained ethnographer from the region, cites a report from 1876 that colonists could not penetrate the territory of the Mbyá without their permission: "The great cacique of the Buha [Mbyá] lives in Baibera (Mba'e Verá), west of the Monday [River] His name is Arapotiiu [Ara Potýju] and he is 30 years old. One cannot go into his capital without his authorization" (Smith & Fogel, 1982, p. 21). Nevertheless, it should be noted that these self-professed leaders had influence only as long as they had popular support among the dispersed Mbyá. A *cacique* might find himself without a following as allegiance shifted to a new leader or dissatisfied followers simply moved away (Müller, 1937/1987, p. 57).

By the end of the 19th century, the distinction between Mbyá, Avá, and Paĩ was evident in the historical records and the ethnographic accounts. Ethnic differences came to be explained in a continuum of acculturation. Most importantly, Avá-Guaraní (Xiripa or Chiripa) were reported to be the most "acculturated." They were seen to have assumed houses, clothes, and market attitudes of the Paraguayan mestizo (Cadogan, 1959). Mbyá, in contrast, were reported to be the least acculturated, closer to the Ache, preferring to avoid contact with the larger society and more dependent on hunting than swidden agriculture. Paĩ to the north were thought to be intermediate between the Avá and Mbyá.

Throughout the 20th century, Mbyá retained a political structure in which regional leaders asserted nominal authority over a number of the dispersed groups. Three assumed regional prominence: Angelo Garay and Pablo Vera in the southern forests and Marcelino Montania in the more northern areas (Cadogan, 1953). These leaders became, for both the Mbyá and in the larger society, the singular point of contact between the state and the small, dispersed and often moving local groups.

Government agencies, such as the *Departamento de Asuntos Indígenas* (DAI), were created to promote the assimilation of Guaraní into the mass of mestizo peasantry (Harder Horst, 2007). In 1950, Cadogan was hired by the government to undertake a Mbyá census, primarily because of his personal connections with the group. Despite having working relations with several of the regional leaders influential with the Mbyá, his work stimulated rumors among the Mbyá that the census was designed to forcibly reduce all Mbyá to a single reservation. Garay and Vera both rejected the census. They wielded sufficient influence to turn the Mbyá population against the effort and demand that he stop the enumeration. Cadogan was forced to abandon the effort (Cadogan, 1953, p. 85).

By 1970, however, a new economy had taken hold in the rural forests. Logging and yerba collection gave way to cattle ranching and intensive farming. Working against land privatization and state policies of assimilation, secular and religious agencies organized to protect indigenous land and society. Several of these groups made direct overtures to the Mbyá. Proyecto Marandú was organized in 1974 by a group of progressive *indigenistas* to provide immediate assistance to indigenous groups, raise consciousness about their perilous circumstances, and most importantly, create a national network of indigenous people. By the time the Stroessner government closed Marandú in 1976, the project had spawned the Consejo Indígena del Paraguay which was recognized by the government as an ethnic federation called *Asociación de Parcialidades Indígenas* (API). Nevertheless, with the same stance they had taken against the government, most Mbyá refused to take part in the developing pan-ethnic federation.

Two other non-governmental groups tried to engage with the Mbyá in the early 1970s. Proyecto Paĩ-tavyterá (PPT) and Projecto Guaraní were organized with the collaboration of the Asociación Indigenista del Paraguay (AIP) to provide health care and legal assistance in land claims for the Guaraní. Like the state, the agencies had little success in establishing working relations with the Mbyá communities. After eight years, only Marcelino Montania would collaborate with Proyecto Guaraní to secure land for his community. The program was only able to secure land for one other Mbyá community in the far north, in the Department of Concepción (AIP & Misión de Amistad, 1977).

Most Mbyá leaders opposed attempts to establish Mbyá *colonias*. Angelo Garay, speaking for the communities of Caazapá, expressed a moral opposition to the titling of land. The Mbyá requested that the land measurement teams pack up their equipment and cease the work, arguing that paper, (titles, censuses, identity papers, and land requests) would only be used against the interests of the forest Mbyá (Fogel, 1997, p. 152). Thus, as soybeans and ranching moved into the forests, the vast majority of the Mbyá remained on private lands that were of increasingly inadequate for hunting, fishing, and their gardens (Wicker, 1982).

Avá communities, unlike the Mbyá, availed themselves of the efforts by *indigenistas,* missionaries, and anthropologists to census communities and secure land. Proyecto Guaraní had considerable success in identifying and organizing Avá-Guaraní communities. When Marandu called for a federation of indigenous peoples, the Avá sent representatives. Several developed into high-profile advocates for their communities. Avá actively sought land and titles and were willing to provide assistance in the identification of their communities. Proyecto Guaraní enumerated the residents, mapped the land, identified leaders, and was able to quickly secure recognition of land claims. By 1977, AIP reported that 50% of the Avá-Guaraní were living on lands that were recognized by the government as *Colonias Indígenas* (AIP and Mision de Amistad, 1977, pp. 90–92).

The historical numbers used to describe the indigenous population are telling of Mbyá opposition to state engagement, and the subsequent inability to protect themselves from the destruction of their forests. In 1950, Cadogan estimated that there were a total of only 2,759 Mbyá (Cadogan, 1953, p. 92). The first national population census in 1980, largely rejected by the Mbyá, found an estimated population of 5,500 and with only 690 families and 2,460 people actually counted (INDI, 1981, p. 41). By 1990, the number of Mbyá that had allowed themselves to be surveyed had only risen to 4,744. In 2002, however, when forest destruction had made it harder to avoid the state, and the enumerators made an all-out effort, the indigenous census counted 14,324 Mbyá (DGEEC, 2002, p. 21). By 2012, the official number of Mbyá reached 21,422, making them the largest of the 17 indigenous ethnic groups in Paraguay (DGEEC, 2012, p. 22). Even after factoring in the high fecundity rate that the Mbyá have shown in recent years, the almost 800% increase represents a dramatic exposure of the previously "invisible" Mbyá, to the nation-state.

INDIGENOUS COMMUNITIES AND DEFORESTATION

When I first began research among the Avá in 1980, I walked 43 kilometers through the forest to the Guaraní communities, which were sheltered by a seemingly endless canopy of trees. Over the decades, the trail has been replaced by an ox-cart track, and then a dirt road; now an asphalt highway cuts into the region. Mestizo colonists cut small fields along the road and the narrow band of crops was expanded, year after year after year, until the forests were destroyed. These northern forests had early incursions by the state and the process of destruction was incremental. By the time the soybean boom hit the region, over 60% of the Avá communities and 80% of the Avá population have been granted land by the state (DGEEC, 2002).

But the southern forests had relatively little clearing until recently, and then it happened with violent speed and disastrous results. When massive land clearing began in the 1970s, only a small number of Mbyá communities had been identified and received any assurances to land. The remaining groups were too small, too isolated or resistant to official contact. Exacerbating the human costs, the most devastating land clearing proceeded in those areas where land control by indigenous residents was the most tenuous.

In short, it was not simply contact with the national society that disrupted these communities, but pace of loss of critical subsistence resources. In the southern forests of the Mbyá, beset by environmental devastation, they were deterritorialized without being displaced.

Two case studies serve to illustrate the process by which Mbyá lost their land in eastern Paraguay. These show that Mbyá face directly the force of neoliberal policies of the state without compromising their goal of life in the forest. Forced from their forest redoubt, they do not capitulate, but confront state agencies directly and forcefully. Even as they are deterritorialized, they assert new ways of resisting the state and redefining what it means to be Mbyá.

Sommerfeld

The isolated Mbyá of the Department of Caaguazú, 200 kilometers east of Asunción, were exposed to the brunt of Paraguay's agricultural development when the Pan-American Highway was cut through the forests in the 1960s. Mennonite families arrived from Paraguay's Chaco, purchased 44,000 hectares, named it Sommerfeld, and proceeded to clear land with devastating speed for commercial agriculture.

Although the purchasers stipulated that the land be uninhabited, the farmers of Sommerfeld encountered three distinct groups of Mbyá. In the first years of the settlement, Mbyá found work in Mennonites' fields, but the massive clearing program soon dislocated their gardens and homes. Mbyá residents were forced onto unused parcels of nonarable land and warned that they would inevitably have to move from what was now private property. As early as 1968, Mbyá complained to the national indigenous peoples agency, the Departamento Asuntos Indígenas (DAI), of Mennonite farmers harassing their villages and attempting to force them off their land. DAI ordered local police to investigate – a request that went nowhere (Harder Horst, 2007, p. 68).

Since that time, the state has been unpredictable in its actions and unwilling to defend the Mbyá from Mennonite aggression. In 1981, the "Estatuto Indígena 904/81" permitted indigenous communities to hold title to land in common, and the Mbyá of the region used the legislation to assert legal right to the land they occupied, demanding that the state recognize their status as one of the *Comunidades Indígenas Nacionales*. The Mennonites responded with direct threats against the beleaguered communities and solicited local authorities' help in removing the families. In 1984, after 16 years of struggle, the Mbyá convinced the new incarnation of the DAI, the Instituto Naciónal del Indígena (INDI), to survey their land that had been claimed by Mennonites. INDI did so and prohibited the farmers from further harassment of the Mbyá. Ignoring the injunction, the Mennonite community stepped up violent reprisals against the local community. Mbyá lands were cleared, cattle were pastured in their gardens, and their access to water was cut off. Women were attacked and men were threatened (DIM, 1987). Soon, the Mbyá villages were laid bare to the sun and surrounded by soybean fields; the residents were hostages on their own land.

INDI quickly reversed its decision to expropriate land for the Mbyá of Sommerfeld. The Mennonite community, epitomizing the independent, capitalist agriculture that the government seeks to promote, won the support of the Ministry of Agriculture and Ranching. INDI agreed to remove 100 Mbyá families onto a reservation 40 kilometers away and claimed they would have 1,000 hectares for their use.

The Mbyá of the region refused to relocate. They appealed for popular support through newspapers and the Catholic Church. With national (and growing international) attention, INDI dared not contradict the "ley indígena" by relocating the Mbyá out of their homelands. After four more years of legal battles, INDI reversed itself yet again in 1989 and expropriated 1,400 hectares from the Mennonites. Nevertheless, the conflict continues. Most of the land they occupy is low and wet, not suitable for gardening. The upland forest has been cleared and houses exposed in the bare landscape. Although many of these communities had space for small gardens, they lost use of the surrounding forests for hunting and gathering.

Isolated from the capital city and cut off from government bureaucrats who signed the document legitimizing their settlement, Mbyá continue to suffer the threats of commercial farmers, who now seek to plant soy on Mbyá land. In a process that is becoming increasingly common in eastern Paraguay, when they refuse to rent land to soybean producers they are charged with standing in the path of national progress; when they do rent, they diminish their own claim to the land. Either way, Mbyá face expropriation by corrupt officials who increasingly settle land conflicts in favor of soy growers (Kelley, 2013).

Proyecto Caazapá

As the soybean miracle took its toll in Caaguazú, new areas of conflict and destruction have been opened to the southeast. Beginning as early as the 1960s, the low forested hills of the Cordillera de San Rafael were beset by integrated rural development projects from two directions. From the west, a series of major development projects were funded by multilateral banks to carry Paraguay's *marcha hacia el Este* to the western flank of the Cordillera de San Rafael. A second wave of development, spurred by private capital, moved into the low hills from the *frontera* with Argentina and Brazil. The forests between the two frontiers became the last refuge of the Mbyá.

From the east, the state-sponsored *Plan De Trigo* and World Bank Project 1418 were funded to improve mechanized agricultural production, develop infrastructure, and promote the sustainable farming, as a means to reduce Paraguay's dependence on wheat imports. The most extensive of these program, Projecto Caazapá became a vehicle by which 381,600 hectares of forest were cleared for agriculture by large corporate farms. The region had been purchased from the government in the 1960s and held on speculation or logged for valuable hardwoods used for furniture and floors (Parquet, 1985, p. 114). At a cost of 54 million dollars, principally from World Bank loans, the Paraguayan government hoped to promote colonization and increase agricultural production in what was considered an under-utilized area. The integrated development project included over 400 kilometers of road to move people in and production out, as well as credit and agricultural extension services for farmers.

Original plans for the project ignored indigenous inhabitants of the region. Estimates in 1979 by the consultants for the project suggested there were fewer than 10 Mbyá families whose nomadic lifestyle made them difficult to locate and identify (Smith & Fogel, 1982). It took the Paraguayan government until 1981 to initiate a census of the indigenous population within the region. Later censuses by indigenistas identified 17 communities and over 730 families, approximately 3,800 Mbyá and Ache (DGEEC, 2002).

The problem of securing lands, however, was more than an issue of identifying and registering Mbyá communities. Most Mbyá refused to cooperate with censuses and efforts to title lands to their communities. More than concerned that enumeration would make them vulnerable to government control, Mbyá leaders in the region refused to accept the legitimacy of the concept of land titling. (Notable exceptions stand out; Cacique Ñaro, when faced with massive land clearing, changed his position and actively sought to engage with the government [John Renshaw, personal communication, December 11, 2014].)

In the meantime, loans were dispersed and roads were built. Private logging trails were quickly cut into the densest forests, traversing the lands used by the Mbyá and often entering Mbyá hamlets. Colonists followed the logging trucks. Paraguayan *latifundistas* sold parcels to peasants. Brazilian entrepreneurs bought land, subdivided it, and contracted their own countrymen to relocate as laborers. Mbyá communities were besieged, their land was cleared, cattle invaded their gardens, and sport hunters killed the game (Prieto, 1991, p. 37). In some cases, Mbyá were evicted, their houses burned, and the residents threatened and brutalized (Reed, 1995, p. 205).

Incongruously, the ecological devastation wrought by Proyecto Caazapá was exacerbated by efforts to title land to indigenous residents. The Mbyá had legal rights to lands and World Bank policies mandated that those claims be recognized. Nevertheless, loggers were aware of the impending demarcation and worked quickly to exploit timber from Mbyá land. By the time, land rights were recognized, the canopy had already been removed.

Attention to the plight of the Mbyá forced the World Bank to suspend its loan to Paraguay in 1987, but what appeared a strong stand against the government quickly proved ineffectual and half-hearted. Loans were suspended after a major portion of the monies had been dispersed and the government was able to continue the project for months with available funds. The Paraguayan government quickly passed legislation (1372/88) reiterating the Estatuto Indígena (904/81) and setting out steps for land titling and detailing communities in need of land. In response to this legislative maneuver and without action to secure Mbyá lands, the World Bank quickly reinstated the loan.

As loggers and farmers were assaulting the forests on the western regions of the San Rafael, roads were cut linking the fertile eastern lowlands along the Paraná with markets in Encarnación and Ciudad de Este. In the 1980s a highway was cut south along the Parana. Brazilian and Japanese investors bought land, cleared forest, and settled immigrant colonists. The Inter-American Development Bank funded the cutting and the paving of feeder roads, turning the fertile soils into an agricultural powerhouse. The Mbyá were removed, often forcibly, onto small tracts between soybean fields. Those that appealed for government defense had their houses destroyed and their forests felled before INDI titles were granted, and often before their emissaries could make a *denuncia* in Asunción (Reed, 1995, p. 206). Mbyá retreated into the last vestiges of forest in higher reaches of the hills.

In 1992, 73,000 hectares of the Cordillera was designated a protected area and in 2002, with direct World Bank support, was placed under a national plan for the management of natural resources. The label and project plan came with few funds, however, so the government was not able to gain control of private lands within the region's boundaries. The region continues to be vulnerable to landowners in search of quick profits.

When Mbyá lands finally began to be surveyed, they were far less than necessary for the indigenous communities. Indigenistas estimated that of the six communities for which data is most complete, 40,000 hectares are necessary to protect them from economic and social dislocation (Rehnfeldt, 1989). Rather than satisfy the World Bank mandate to protect indigenous peoples, the government only provided the communities with land sufficient for a life as peasant farming, as provided in the national indigenous land law. Some Mbyá communities refused to accept both the premise and the outcome of this land program. By 2002, land had been acquired for only 30 of the 53 Mbyá communities of the Departments of Itapúa and Caazapá, leaving over 40% of the population, almost 1,400 people, without land (DGEEC, 2002). The Guaraní and indigenistas continue to fight for the remaining land, even as it is settled and cleared for soybean fields and cattle ranches. Of the total Paraguayan Mbyá population of 14,537 identified in the national census of 2002, only 6,316 (43%) had been provided land. The vast majority of these hamlets were provided only small parcels suitable for house lots (DGEEC, 2002).

INDIGENOUS ENVIRONMENTAL REFUGEES

The Mbyá take refuge in Asuncion because their lands have been cleared, plowed, and planted for massive, mechanized agro-industrial production. The forest that once sheltered these communities is gone, the game is killed off, and the fruits of the forest are replaced by soybean fields and ranches. As the last corners of the forest are cleared for planting, Mbyá are run off their land by threats, legal action, and physical violence. The last sanctuary for these displaced people is the most incongruous; they turn to the anonymity of the city, where civil violence and state power are tempered, and where the urban political structure lacks the capacity to exclude them. However, as the Mbyá seek to establish themselves in the city, they do so without the forest environment from which they have drawn material needs and social identity. Deterritorialized in the most concrete way, they are cast adrift in the urban society and space. They are quite literally, people of the forest without forests.

Mbyá-Guaraní collect on street corners, empty lots, and along storm water drainage canals outside the city center. In small groups of 5–15, families search out urban space where they will not be disturbed by landowners or shopkeepers. These spots lack water, buildings, or services, but are chosen for their ability to shelter the small groups in tall grass or trees, yet remain close to streets and markets where they might get access to food or money.

During the day, Mbyá filter through the city to beg, or work, or scavenge. Many find their way to the inter-urban bus station, the nexus of the various networks of rural "micros" that bring rural dwellers to the city. The scene is a tumult of noise as travelers search for ticket offices and platforms and merchants hawk food and wares. Mbyá women with babies stand at the edge of the property, imploring travelers with a word or a look for a bit of change. Young Mbyá men run between the micros, selling soda and gum for a few cents profit on each. Young Mbyá women find work in nearby bars serving drinks and, one suspects, having sex with the transient men who filter through.

As one explores the urban world of the Mbyá in the city, it becomes progressively darker. Adolescents are especially vulnerable to the problems of snorting glue, sex work, and violence. Abandoning their parents and younger siblings, these *niños de la calle* form bands among themselves and are lost to families and communities, soon consumed by drugs and disease. They spend their days sleeping and inhaling small plastic bags of "*cola de zapatero*," shoemakers glue which contains toluene, a neurotoxin. They sell sodas on city buses, sex on the street, or panhandle for the money. With the cost of a bag of glue only Gs. 1,000 – about a quarter of a US dollar, it is easy to stay high most of the day. They are the most extreme casualties of the disaster that has befallen the Mbyá.

My work has been with Guaraní in rural areas and specifically in the forest. Despite noting the presence of Mbyá in city streets, I had little knowledge of their plight until my own Guaraní friends began showing up in Asuncion in 2011. Working with these Guaraní and local anthropologists in 2011 and 2013, I was able to get to know a number of urban refugees, and explore their story of dislocation and struggle. Several of these cases deserve discussion here. Three are Mbyá settlements, and the fourth is a recognized *Comunidad Indigena Nacional* that has been organized by Avá settlers near the city garbage dump.

Kilometro Doce

Kilometro Doce is a dusty, broad, and busy intersection 12 kilometers outside the city center, in the town of Luque, behind the luxurious Hotel Bourbon. Luque was a small community that is fast becoming a middle-class suburb in the larger metropolis. Farmers' fields of previous generations are giving way to warehouses and shopping centers. A survey carried out in 2011 found that there were 23ty-three Mbyá families, close to 100 people, in the area of Kilometro Doce (Grupo Sunu de Acción Intercultural, 2012). With few houses nearby, and a variety of undeveloped lots being held on speculation for future development, the place is well suited for small encampments of Mbyá to fade into invisibility within a few paces of busy urban world.

On one visit to Kilometro Doce in 2013, I found a small ragtag group of nine Mbyá lounging on the curb of the street, inches from speeding traffic. They were eating out of a Styrofoam clamshell, food which had been purchased from the Supermercado MercaMundo across the street. Plastic bottles, filled with water sat on the sidewalk beside them. At first impressions, the scene seemed oddly similar to meals in the forest, adults squatting around and eating out of a single cooking pot, kids leaning over their shoulder to grab a bite or a bone. Here, however, the tiny group was incongruously embedded in a world of asphalt and exhaust.

They looked up with interest at my approach. The single adult man straightened up, fork in hand, and nodded his head. It wasn't just that I was walking toward them; my eye contact with the members of the group set me off from others hurrying passed in cars or on foot. It was clear that the invisibility of the little group depended not so much on their physical unobtrusiveness, as the fact that they are not seen in the urban context.

Most casual conversations with urban refugees followed a pattern that expressed a great deal with few words. My simple questions won answers that both said nothing and everything. "We came from Caaguazu." "We just got here." "There is nothing there for us." And most Mbyá retained the hope and expectation that they would return to their communities and regain their lives. When asked what they intended to do, the only plan most would formulate was, "We are going to go back." This it seemed was both bitter acceptance of the intractability of the problems and defiant statement that things would change for the better.

In fact, most groups regularly return to their rural homesteads. There is a seemingly continual flow of people between the impoverished rural communities and the city streets. Small groups make their way to the bus station and buy passage to the hinterland, rejoining friends and family on the small plots that they hold onto in the countryside. Others make the return trip, eventually reuniting with community members on the street.

When these adults leave the street at night, they find refuge in overgrown land nearby. Here they are protected from public view during the day, and almost impossible to find after dark. Nights are Asunción may be cold, but squatters don't build fires. With nothing to cook and only a few eucalyptus branches to burn, fires are more trouble than worth – and might attract attention.

Young people seem to formed their own band, affiliated with yet independent from that of adults. At Kilometro Doce they hang out beside an urban drainage ditch, an area they have fashioned as a kind of *pied-à-terre* out of sight of the passing traffic. The lot has clearly being used by a number of people and for a while. Paths have been worn through the dry grass. Toward the back of the lot, piles of human waste and discarded glue bags line the paths; torn clothes and empty carry-out containers are tossed into the brush; abandoned blankets mark the spots where the little groups sleep.

Terminal

Many of the small Mbyá groups pass through the region of the city known as "*Terminal*," the inter-regional bus station outside the city center. Here, on one visit, I found a Mbyá mother with a babe-in-arms, on a small grassy knoll outside the busy building. Several small children played on the sidewalk in front of her. She stood with an older daughter, who was dressed in a clean knit blouse and black polyester slacks. They surveyed the traffic with almost studied boredom. I strike up a conversation with innocuous questions: "Where do you come from?"; "How long have you been here?"; "Is that your baby?" I finally ask the question I'm most interested in: "Where are you staying?" She answers, "Over there, down that street." I motion. She nods assent. I wait a minute and head off, taking leave as I would in the forest, with a simple statement, "I have gone."

The cobble street is strewn with the litter left by transient people: food wrappers, old bus tickets and plastic shopping bags. I pick my way down the street and the organized chaos of the bus station give way to the disorder of *pensiones* and *loncherias* and prostitutes and truck drivers that surround travel centers.

Mbyá often rent rooms in the shabbiest of the pensiones. Dark, narrow hallways off the dirty street open into small dirty cement courtyards. After being on the street, they find some respite in a calmer surroundings, resting and sharing share tereré with friends, family, and compatriots. The rooms tend to be small and dark, with plank wood doors and ancient metal clasps. Locks are unimportant as there are few furnishings or belongings. Without beds, the rooms have a jumble of bedding on the floor, usually wrapped around some somnolent figure. They were lit only with the scant light from the open door. These rooms were only big enough for a couple of families, and rent was paid by the day – or hour – so there is considerable pressure for each person to come up with some cash daily, albeit a small sum.

Mbyá sometimes join forces to rent small, shabby houses in the residential neighborhoods nearby. One small non-descript house can serve as an example. Although the house looks empty on the face of it, with a closed door and shuttered windows, in the back is a hubbub of kids and women and babies spilling out to the small dirt-covered courtyard. Six "families" share the space, chipping in Gs 100,000 (about US$25) per month to cover the rent. They make money in the street, mostly begging, to pay rent and buy what little they need. The landlord, if they are paying the owner, is obviously collecting the rent without offering any maintenance. Like their compatriots in pensiones and on the street, these Mbyá come from Ca'aguazu. Most have been here only a matter of a few weeks and plan to return to the rural area soon.

Plaza Uruguaya

The Mbyá exist in a suspended state in the urban environment. They engage with urban citizens, yet they are not urban citizens. Whether squatting in a vacant lot, selling sodas in city buses or begging at the bus terminal, they survive day-to-day. This is not to suggest that they do not establish new social relations and identities in their new environment. In fact, Mbyá have organized high profile, and contentious demonstrations in opposition to the state and for the return of their forests.

As the soybean boom moved through central Paraguay in the 1980s and 1990s, Mbyá leaders often made the trek into Asunción to denounce the usurpation of their land and the clearing of their forests. Appearing at the offices of the INDI, they protested their losses and appealed to the bureaucracy to secure their lands or make restitution. Usually they were ignored or went back to the campo with promises that offered little help in the violence of rural struggle.

Actions taken over the Easter weekend in 2007 mobilized Mbyá refugees in a very public confrontation. Rather than settle for empty assurances, leaders amassed displaced Mbyá to occupy Plaza Uruguaya, one of the largest plazas in the city center. Unlike other, more well-groomed plazas in the city center, the Plaza Uruguaya is of the common people. The worn benches, over-used lawns, and dry fountains sit a bit forlornly next to the abandoned train station. The Mbyá carved out space to site among the workers, street vendors, and occasional street walkers, with a public water tap and restroom nearby.

Under nominal direction of self-proclaimed leaders pushing land claims, Plaza Uruguaya soon turned into a veritable tent city for the Mbyá, with bright plastic tarps strung over rope and families warming themselves beside small cook fires with water for yerba mate. Newspapers and television crews took notice and the encampment soon attracted both its defenders and critics (e.g., Indigenas Mbyá, 2007). Under Stroessner there had been little political space for public demonstrations, but the Mbyá took advantage of liberal reforms and government inefficiency to press their case.

The settlement developed into an extended occupation that kept the Mbyá land conflict in the public eye and on the government agenda.[2] The Mbyá did not acquiesce, INDI had little power to satisfy their demands, and the progressive government of Lugo refused to commit itself. The half-hearted efforts to solve the problem allowed the occupation to drag on for four years. In 2011 when a conservative party took over the municipal government, it finally removed the Mbyá *carperos* from the plaza and erected a wrought iron fence around the space. The Mbyá removed themselves to the abandoned train station across the street.

Mbyá continue to demonstrate at the INDI offices. Restricted from occupying the first floor of the public offices, they set themselves up on the sidewalk outside. They arrive from far flung corners of the city each morning when offices open and remain on the sidewalk outside, lounging and talking and caring for their children, throughout the day. At the end of the day and on weekends they disperse to vacant lots, rented rooms, and abandoned buildings.

Mbyá demonstrations have the singular purpose of acquiring sufficient uncleared land to permit the Mbyá to return to the forests. From the first, the Mbyá demanded large forested parcels and were satisfied with nothing less. Few areas of eastern Paraguay remain of sufficient size and quality to permit indigenous groups to return to a semi-nomadic hunting lifestyle, and INDI had neither the power nor funds to satisfy Mbyá demands. The process was as frustrating for the Mbyá as it was exasperating for government officials. Representatives for both groups would visit possible parcels and ultimately some or all of the Mbyá would reject the land as too small or lacking good cover for hunting or gardens.

In many ways, the Mbyá reaction to the urban environment suggests that, however hard their begging or abuse or invisibility in Asunción is, it is preferable to their existence on the parched hot parcels that remain for them in the countryside. The moment-to-moment existence on the urban street is better than the future-less life that they face in rural areas. The city remains the last redoubt against the structural violence of the state.

In addition, however, Mbyá wield their presence in the city as a challenge to state authority. They carry the conflict of their past into the present and from their rural homes into the heart of Asunción. The Mbyá develop neither the plan nor the tools to engage effectively in the urban sector. Without the intention to establish a secure footing in the city, they do not acquire the education, develop the political training or learn the work skills to carve out a niche for themselves. When they do stand in opposition to the government, they do so without short-term objectives in mind, but in a tenacious effort to achieve their long-term goals, a return to independence and autonomy in the forests. In the end, the public struggle for land might be understood as a means to publicly express Mbyá resistance to the environmental destruction that had been visited upon their communities.

Cerro Poty

Other Paraguayan indigenous groups have suffered the effects of land loss, and even the predations of the soybean miracle, and not ended up on the streets. Many Avá, also hard-pressed by soybean farmers, have recently moved from rural areas and settled in urban enclaves. In contrast to the Mbyá, most Avá find work, build homes, and in some cases win international aid and government recognition as *Comunidades Indígenas Nacionales* in the city. One case deserves special mention.

One group of Guarani, primarily Avá, has collected along the river near the municipal dump "Cateura" in the settlement called "Cerro Potý" ("Flower of the Hill"). Originally attracted to the proximity of the refuse for foraging, the group has grown to over 300 people, with a preponderance of women and children, making up some 60 native families.

The Asunción city center is bounded on the west by the Paraguay River, and Argentina on the far side of the water. The suburbs spread over the highlands to the east. Squatters looking for a place to build shelter must move onto unclaimed land on muddy shores of the slow moving river. The first Avá families came in 1990 from the Department of Guairá. Undisturbed on unoccupied marshy lowlands near the city dump, they quietly constructed houses from scavenged material and adopted a sense of stability in the urban environment. Friends and relatives joined them from other desperate Guarani communities, gravitating to this "safe zone" along the river.

Several of the first families at Cerro Poty arrived from the Guaraní community Agua'e, in the Department of Canindeyu. Despite that Agua'e was one of the earlier communities to have its land recognized by the state, they had only 980 hectares. By 2009, there were 80 families, with about 420 individuals – allowing an average of only 12 hectares per family. A paved road was constructed linking the area with the Brazilian border and the surrounding forests were cleared and soy planted to the very borders of the *colonia*. Soon thereafter, community members began to report that crop dusting of herbicides and pesticides was applied without attention to winds, natural barriers, or the proximity to houses and schools. Chemicals drifted into the *colonia* school sickened domestic animals and fouled the community's water; residents suffered diarrhea, fever, and nausea (López, 2015). Families abandoned their land and moved toward Asunción, finding their way to Cerro Potý.

The 2002 national census counted 90 Guarani at Cerro Poty; in 2008 there were 135, most under 17 years of age; and by 2013 there were 53 families, probably 265 people. In addition, two other settlements sprang up nearby, poetically called Cateura 1 and Cateura 2, after the nearby dump. Interestingly, although some of the early migrants were Mbyá from Caaguazu and Guaira, these families soon filtered away to live elsewhere.

The location of Cerro Potý in los Bañados del Sur is convenient to the most important source for income: the landfill for scavenging. Despite over-flowing its banks and spilling into the marshy river, the refuse continues to accumulate at a rate of over 1,200 tons per day, or a kilo per person in the city. In the midst of the flies, dust, toxic effluent, and flooding, Guaraní can earn a living by recycling a wide variety of materials. The lucky forager can find lead, copper, or construction materials. But even without these, Guaraní can gather plastic bottles to wash for resale. On a good day, residents earn as much as eight dollars, allowing a family to provide for basic food needs – if little else.

More than earning cash, the people of Cerro Potý have organized as a community, solicited government recognition and accepted considerable public assistance in their efforts to establish a community. Refugees from various Avá communities assembled near the landfill, elected a leader and won INDI certification as a *Comunidad Indígena Nacional*. With government assistance, the group of families have built a primary school, erected a religious center and a community meeting area. As the public took notice of indigenous peoples in the landfill, the group effectively managed television, newspapers, and social media to make their case. They attracted international assistance, including the construction of raised wood houses to avoid the toxic effluent that drains through the region during heavy rains.

The institutional infrastructure of Cerro Potý did not appear sui generis, it is the result of decades of institutional development in the residents' rural communities. Agua'e was one of the very first communities to work with indigenistas of Projecto Guaraní. The community was recognized and receives land assurances with the first wave of grants by INDI. Leaders were regular participants in regional and national ethnic federations. In 1978, Agua'e was chosen as the site of a Guaraní health center, where native peoples could find treatment without traveling to Asunción. By the time soybean farmers drove the people off their land, they had decades of experience in working with and fighting against the larger system. It is not by chance that the current leader of Cerro Poty is the daughter of the elderly cacique of the community they left.

In short, Cerro Potý stands as one example of the success that Avá-Guaraní have had in establishing stable infrastructure in the urban arena. Although fraught with problems (sickness and pollution, to name the obvious), they draw on their rural experience to articulate effectively with the state and the Paraguayan nation, winning a modicum of stability in the urban system.

CONCLUSIONS

How do we understand the effect of environmental destruction on the Mbyá? Recent anthropology calls attention to the importance of place in social identity and the dynamic process by which historical relations structure urban migration. Environmental destruction engendered by soybean production has displaced the Mbyá from the forests, disrupting not just of their subsistence, but an identity, a structure of social relations and a relation to the Paraguayan state. They arrive in the capital city without social networks, the technical skills, or cultural understandings to become part of the urban society.

But, deterritorialization implies reterritorialization, and in the case of displaced indigenous peoples this involves struggle with the state not simply over resources, but over the social relations that integrate the group into the dominant society. Historically, the Mbyá have taken a strong and strikingly resistant posture with relation to the state and the national society. They defended themselves from state encroachment in their forests for centuries and, when the cattle ranchers and soybean farmers arrived, Mbyá refused both assimilation and state assistance. They have been intransigent in their demands for the space and resources to maintain an independent and autonomous region.

Forced into the city as refugees, they continue their historical resistance to the state. Mbyá-Guaraní arrive in the city center and gravitate to friends and relatives already there. But rather than make their way toward Cerro Potý, or any of the other recognized indigenous enclaves in the city, Mbyá tend to end up in small groups on the street, in various plazas or in houses or rooms rented to transients near the city bus station. They rarely avail themselves of city services, nor do they seek employment, health, and educational benefits that can be found in the city. Mbyá seek out the spaces that are at once obvious and invisible, engaging in activities that provide immediate sustenance without developing a stable life in the city.

How do we explain the experience of Mbyá Guaraní groups in Asunción? How can we understand the process by which some groups establish themselves, and perhaps prosper, in the urban environment and others fail to find a foothold and never quite achieve stability.

It would be easy to write the Mbyá off as passive victims of their refugee status, inept and impotent in the face of the expanding Paraguayan state. They seem to conform rather easily to the concept of the hypermarginalized minority. However, to see the Mbyá simply as a disenfranchised population renders invisible the agency that Mbyá exhibit in the face of the state. The Mbyá refusal to engage with the urban system can be understood as a means of resistance. It is important to look beyond immediate environmental destruction to understand the historical relation of the Mbyá with the state. Over four centuries the Mbyá avoided missionaries; held the yerba economy at arm's length; refused to be enumerated or mapped by the government; and rejected the overtures of indigenistas. They have been remarkably successful. In many cases their communities were unknown to the state until their residents began to show up on roadsides, displaced from their forests by private landholders and deforestation. In the city, Mbyá continue this stance of disengagement with state structures, remaining invisible to agencies and government entities alike. They do not establish social infrastructure that might serve as support for permanent, or even stable, engagements with the city.

Even in their protests for land and their occupation of the Plaza Uruguaya, the Mbyá exhibit a intransigence in their negotiations with the state. They refuse to compromise; they accept neither the plots of land the government offers, nor the option of becoming an urban enclave. They hold out for forests that would allow them to maintain an independent and autonomous lifestyle, an option that is less common in the experience of Mbyá as the forests become a memory.

In this, we see the Mbyá engaging in ongoing resistance to assimilation and integration, based in their historical opposition to the state. Historical practices take new meanings in new contexts. Standing on the street, camping in abandoned lots, begging at traffic intersections are the desperate acts of a disenfranchised minority, but more than that, Mbyá are engaging in social resistance in a manner not unlike Scott's (Scott, 1987) "weapons of the weak." In explicit rejections of the middle-class norms of the city, they are refusing to recognize the power of state domination.

The case of the Mbyá has important lessons for our understanding of indigenous struggles in the fluid context of the state-sponsored environmental destruction. Facing ecological disaster and social approbation, the Mbyá seek a means to negotiate a new way of being indigenous in the contemporary physical, political, and social environment. Offered a choice between assimilation and integration into the urban society on one hand and the hypermarginality of living as displaced refugees on the other, they choose a continued and permanent state of conflict within the urban sector. They camp in empty lots and abandoned buildings and, finally, claim the city plaza as a means to confront the state on its own territory. With tactics drawn from those of previous decades, and adapted to the current political climate, they fashion a new way of being Mbyá, without sacrificing the old.

NOTES

1. To its credit, in 2006 the national government passed landmark forestry protection law. This, in conjunction with the fact that valuable little timber remained, dramatically reduced the rate of forest destruction.
2. For an analysis of Mbyá demonstrations in Asunción, please see Reed (2015).

ACKNOWLEDGMENTS

I owe a debt of gratitude to Marilyn Renfeldt, an anthropologist and *indigenista* in Asunción, who shared her considerable experience with Mbyá both in the forest and on the streets of Asunción. The research also owes much to Anki Vargas, who introduced me to many new and old Guaraní friends in the city. Research in Paraguay was carried out in 2011 and 2013 under the auspices of the Sociedad de Estudios Rurales y Cultura Popular.

REFERENCES

Aguirre, J. F. (1949). *Diario*. Buenos Aires: Imprentadora Tucuman.

AIP and Misión de Amistad. (1977). *Población y tierras indígenas en la región oriental de la Republica del Paraguay*. Asunción, Paraguay: Asociación indigenista del Paraguay and Misión de Amistad.

Barnes, J., & Allen, D. (1985). The causes of deforestation in developing countries. *Annals of the Association of American Geographers*, *75*(2), 163–184.

Basso, K. (1996). *Wisdom sits in places: Landscape and language among the Western Apache*. Albuquerque, NM: University of New Mexico Press.

Bessire, L. (2014). The rise of indigenous hypermarginality: Native culture as a neoliberal politics of life. *Current Anthropology*, *55*(3), 276–295.

Cadogan, L. (1953). El Problema de la Población Mbyá-Guaraní del Departmento del Guairá. *Boletín Indigenista*, *11*(1), 72–92.

Cadogan, L. (1959). Como Interpretan los Chiripá (Avá Guaraní) la Danza Ritual. *Revista de Antropología*, *7*, 65–99.

Cadogan, L. (1960). En Turno a la Aculturación de los Mbyá-Guaraní del Guaira. *America Indigena*, *20*(2), 133–147.

Cadogan, L. (1962). *Guerra de Doscientos Años de los Monteses del Taruma*. Yvytyrusu, 15.

Clastres, H. (1995). *The land-without-evil: Tupi-guarani prophetism*. Champaign-Urbana, IL: University of Illinois Press.

DeLeuze, G., & Guatari, F. (1983). *Anti-Oedipus: Capitalism and Schizophrenia*. Minneapolis, MN: University of Minnesota Press.

DGEEC. (2002). *Segundo Censo Nacional Indígena de Población y Viviendas*. Secretaria Tecnica de Planificacion de la Presidencia de La Republica, Dirección General de Estadística, Encuestas y Censos. Asunción, Paraguay.

DGEEC. (2012). *Censo Nacional de Población y Viviendas para Pueblos Indigenas en el Paraguay: Resultados Preliminares 2012*. Dirección General de Estatística, Encuestas y Censos. Asunción, Paraguay.

DIM. (1987). Somerfeld: Continua presión Menonita. *Dialogo Indígena Misionero (DIM)*, *8*(26), 6–9.

FAO. (2010). Global forest resources assessment 2010. Food and Agriculture Organization of the United Nations, Rome.

Fogel, R. (1997). *Mbyá Recové: La resistencia de un pueblo indómito*. Pilar, Paraguay: Universidad Nacional de Pilar.

Gordillo, G. (2011). Longing for elsewhere: Guarani reterritorializations. *Comparative Studies in Society and History*, *53*(4), 855–881.

Grupo Sunu de Acción Intercultural. (2012). Informe Final de Investigación "Validación de un Modelo de Restitución de Derechos de la Niñez y Adolescencia Indígena." *Asunción*. Paraguay: Grupo SUNU de Acción Intercultural.

Gupta, A., & Ferguson, J. (1997). *Culture, power, place*. Durham, NC: Duke.

Harder Horst, R. (2007). *The Stroessner regime and indigenous resistance in Paraguay*. Gainesville, FL: University of Florida Press.

Huang, C., Sunghee, K. S., Townshend, J. R., Altstatt, P., Davis, A. (2009). Assessment of Paraguay's forest cover change using landsat observations. *Global & Planetary Change*, *67*, 1–12.

INDI. (1981). *Censo y estudio de la población indígena del Paraguay – 1981*. Asunción, Paraguay: Instituto Paraguayo del Indígena.

Indígenas Mbyá de Plaza Uruguaya Piden Carpas Para Protegerse del Frio. (2007). *ABC Color*, 2007/ 7/11: 16a.

IPCC. (2007). *Fourth assessment report: Climate change*. Copenhagen, Denmark: Intergovernmental Panel on Climate Change.

Kelley, M. (2013). Yvá Potý Rising: From the Ashes, a Cause for Hope. *Cultural Survival Quarterly*, *37*(1), 16–17.

López, P. (2015, January 7). Fiscalía deja morir causa por fumigaciones ilegales en Campo Agua'e. *E'a*, p. 1.

Markley, B. (2012). *Paraguay: Oil seeds and products annual*. United States Department of Agriculture. Washington, DC: United States Printing Office.

McSweeney, K., & Jokisch, B. (2007). Beyond rainforests: Urbanisation and emigration among lowland indigenous societies in Latin America. *Bulletin of Latin American Research*, 159–180.

Müller, P. F. (1937/1987). *Etnografia de los Guaraní del Alto Parana*. Buenos Aires: Steyler Missionswissenschafteliche Institut.

Nimuendaju, K. (1978). Los Mitos de Creación y de la destrucción del mundo. Lima: Centro Amazónico de Antropología y Aplicación Práctica.

Parquet, R. (1985). *Las empresas transnacionales en la economía del Paraguay*. Unidad Conjunta CEPAL/CET, Buenos Aires. Unpublished manuscript.

Prieto, E. (1991). Entre la resignación y la esperanza. Asuncion, Paraguay: Centro de Estudios Humanitarias (CEDHU).

Reed, R. (1995). *Prophets of Agroforestry*. Austin, TX: University of Texas.

Reed, R. (2015). Mbyá Resistance and the Politics of Space in Asunción. *Urban Anthropology and Studies of Cultural Systems and World Economic Development*, *44*, 1–2.

Rehnfeldt, M. (1989). Los Mbyá y la ocupación del espacio. *Mbya-Guarani*, *1*(2), 3–4.

Scott, J. (1987). *Weapons of the weak: Everyday forms of peasant resistance*. New Haven, CT: Yale.

Smith, R., & Fogel, R. (1982). *The Invisible Guaraní: The Effects of Development Projects on the Chiripá and Mbyá of Paraguay*. Unpublished manuscript.

United Nations, Economic and Social Council. (2010). *Human rights: Implementation of the United Nations declaration on the right of indigenous peoples*. United Nations, Permanent Forum on Indigenous Affairs. New York, NY: United Nations.

Velázquez, A. M. (2013, April 15). *ABC Color Digital*. Los indígenas urbanos reclaman la legalización de 17 asentamientos. Retrieved from http://www.abc.com.py/edicion-impresa/locales/los-indigenas-urbanos-reclaman-la-legalizacion-de-17-asentamientos-560673.html

Wicker, H. (1982). Mbyá, Paĩ-Tavyta und Chiripá en Ostparaguay. *Jahrbuch des Bernischen Historischen*, *61*(62), 101–123.

ABOUT THE AUTHORS

Olivia Barnett-Naghshineh is a PhD Student in Social Anthropology at the University of Auckland, New Zealand, and has recently completed fieldwork in the PNG highlands as part of a Marsden Fund project "Food Security in a Rapidly Urbanizing Country: The Goroka Fresh Food Market, Papua New Guinea," directed by Dr. Mark Busse. She has an MSc in Development Studies from the School of Oriental and African studies (SOAS) and her research interests lie in gender relations, economic anthropology, the political economy of food, and the anthropology of markets and environmental anthropology.

Ana Maria de Souza Mello Bicalho is a Rural Geographer (PhD, University of London) who is an associate professor of the Departamento de Geografia, Universidade Federal do Rio de Janeiro, Brazil, and was chair of the International Geographical Union Commission on the Sustainability of Rural Systems between 2007 and 2012. She has over four decades of research experience in the Amazon, the Central-West, the Northeast, and the Southeast of Brazil and has published articles in *The Journal of Peasant Studies, Geographical Research, Horizons in Geography* and has edited a number of books on sustainable rural development.

Sophie Bouly de Lesdain, previously a Senior Member at Saint Anthony's College, Oxford, is now an Associate Researcher at IIAC-LAU Lab (CNRS-EHESS) and expert researcher at EDF Research and Development Department. She earned her PhD in social and cultural anthropology in 1996 at the Sorbonne, later becoming a lecturer (1999–2004). She has conducted ethnographic research in Cameroun, Senegal, and France. She initially worked on food production, processing and distribution in agricultural societies and on redistribution mechanisms in societies of egalitarian ideology, before moving on to urban manifestations of economic contexts, specifically in relation to the constitution of "exotic" food sectors in France. She has recently been focusing on household energy production and consumption.

Her research interests include economic anthropology, urban anthropology, energy and the environment.

Jane I. Guyer is George Armstrong Kelly Professor in the Department of Anthropology at Johns Hopkins University. She graduated from the London School of Economics (BA Sociology, 1965) and the University of Rochester (Anthropology 1972), and has taught at Harvard, Boston, and Northwestern Universities. At Northwestern University, she was Director of the Program of African Studies (1994–2001). Her field research in Africa has focused on agriculture (*An African Niche Economy. Farming to Feed Ibadan 1968–88* (1997)) and money, which drew her into historical work, such as her edited collection entitled *Feeding African Cities. Studies in Regional Social History* (1987). She published *Marginal Gains. Monetary Transactions in Atlantic Africa* in 2004 (University of Chicago Press) and has another collection of essays forthcoming with the University of Chicago Press. She was elected to the National Academy of Sciences in 2008.

Scott William Hoefle is a cultural and social Anthropologist (DPhil, University of Oxford) who is a full Professor and chair of the Postgraduate Program in Geography of the Departamento de Geografia, Universidade Federal do Rio de Janeiro, Brazil. He has 40 years of research experience in the Amazon, the Northeast, and the Southeast of Brazil and has published articles in *Critique of Anthropology*, *The Journal of Peasant Studies*, *Geographical Research*, and *Géographie et cultures*.

Hao Li holds a PhD in water resources management and an MSc in Regional Economics. He has been a post-doc in the UNESCO-IHE. His doctoral dissertation was about the impact of climate change on water resources security and adaptation measures in China. Currently he is an associate professor in the Changjiang River Scientific Research Institute, of the Changjiang River Commission. His major research fields include climate change, water governance, trans-boundary river management, and integrated water resources management. Related papers have been published in "Water Policy," "Disaster Advance," and "International Journal of Water."

Ty Matejowsky was born in Bossier City, Louisiana, in 1970. He grew up in Brenham, Texas, and earned BA and MA degrees in Anthropology from Texas A&M University in 1993 and 1995, respectively. He received his PhD in Anthropology from Texas A&M University in 2001. He is currently an Associate Professor with the Department of Anthropology at the

University of Central Florida. His ongoing research in the Philippines examines a number of issues related to globalization, including fast food, urban development, disaster responses, and international migration. Publications include "The Incredible, Edible *Balut*: Ethnographic Perspectives on the Philippines' Favorite Liminal Food" (2013) in *Food, Culture and Society*, "Like a 'Whopper Virgin': Anthropological Reflections on Burger King's Controversial Ad Campaign" (2010) in *Studies in Popular Culture*, "Fast Food and Nutritional Perceptions in the Age of 'Globesity:' Perspectives from the Provincial Philippines" (2009) in *Food and Foodways*, "Jolly Dogs and McSpaghetti: Anthropological Reflections on Global/Local Fast Food Competition in the Philippines" (2008) in *Journal of Asia-Pacific Business*, "Convenience Store Pinoy: Sari-Sari, 7-Eleven, and Retail Localization in the Contemporary Philippines" (2008) in *Philippine Quarterly of Culture and Society* and "SPAM and Fast Food 'Glocalization' in the Philippines" (2007) in *Food, Culture and Society*. He can be reached at Ty.Matejowsky@ucf.edu

Patrick B. Patterson is currently a Research Associate in the Population Health Inequities Research Centre in the O'Brien Institute for Public Health at the University of Calgary. He completed his PhD in Social Anthropology at the University of Calgary in 2014. He conducted ethnographic studies of the logging industries in the Okanagan/Shuswap and East Kootenay/Columbia regions in British Columbia, Canada. Since 2004, he has also worked as an analyst on interdisciplinary projects studying the implementation of antitobacco policies in hospital settings, and studying the development of Canadian public policies to address household food insecurity. His interests revolve around interplay between the economic ecologies of industrial societies, social and cultural factors in policy formation, and local concepts of risk and safety.

Victoria C. Ramenzoni, PhD, is currently a contracting Social Science Analyst with the National Oceanic and Atmospheric Administration, USA. She obtained her PhD in environmental anthropology at the University of Georgia. Originally from Argentina, she has carried out research in decision-making about resource use, open access systems, environmental fluctuations, and unsustainable practices in Eastern Indonesia, Latin America, and the United States. She has a background in fishery sciences, economic anthropology, behavioral ecology, and mixed methods. She is interested in investigating how socio-economic and climate related uncertainty shapes decision-making processes and resource use practices.

Richard K. Reed is Professor in the Department of Sociology and Anthropology at Trinity University in San Antonio, Texas. Reed completed his PhD at Harvard University in Social Anthropology, focusing on the historical integration of indigenous Guarani into extractive industries in the forests of Eastern Paraguay. Notable among his books are, *Prophets of Agroforestry: Guarani Communities and Commercial Gathering* (Texas 1995) and *Forest Dwellers, Forest Protectors* (Allyn and Bacon 2008). Recent deforestation for cattle and soybeans has made forest life impossible for many Guarani. The work included here examines the transformation of Guarani into environmental refugees in the capital city, and the implications it has for indigenous identity.

Pasang Yangjee Sherpa is a Lecturer in the department of anthropology at Penn State University. She earned her PhD in anthropology from Washington State University (2012). She is a native Sherpa woman from the Mount Everest region, born and raised in Kathmandu. Her past research (2009–2012) examined Sherpa perceptions of climate change and institutional responses to climate change effects in the Mt. Everest region of Nepal. Her current research interests lie in finding practical solutions to environmental problems by discovering sustainable engagement strategies to increase accessibility to scientific knowledge and institutional resources by people regardless of their socioeconomic background at the local level.

Monica L. Smith is Professor of Anthropology at the University of California at Los Angeles, and serves as the director of the South Asian Laboratory at UCLA's Cotsen Institute of Archaeology. She has directed archaeological research projects in India and Bangladesh, and participated in fieldwork in Italy, Tunisia, Turkey, Egypt, England, Madagascar, and the American Southwest. Smith is the author of *A Prehistory of Ordinary People* (University of Arizona Press, 2010) and co-author (with R.K. Mohanty) of the excavation report *Sisupalgarh* (Indian Archaeological Society, 2008). Her research on archaeological subjects includes the analysis of early state networks, biological models of human behavior, and the role of the past in national identity. At UCLA, she teaches courses on archaeological theory, cognition and material culture, and the development of urbanism.

Meine Pieter van Dijk (PhD Economics Free University Amsterdam) is an Economist and Professor of Water Services Management at UNESCO-IHE Institute for Water Education in Delft, professor of entrepreneurship at Maastricht School of Management (MSM) and professor of Urban management at the Institute of Social Studies (ISS) and at the Institute of Housing and Urban development Studies (IHS) of the Erasmus University in Rotterdam (EUR), all in the Netherlands. He is member of the research schools CERES and SENSE. He worked on and in developing countries since 1973 and as a consultant for NGOs, the Asian Development Bank, the Inter-American Development Bank, the World Bank, different bilateral donors and UN agencies.

Donald C. Wood is an Associate Professor in the Graduate School of Medicine, Akita University, Akita, Japan, where he has worked since completing a doctoral degree in cultural anthropology at the University of Tokyo in 2004. Prior to that, he studied anthropology under Norbert Dannhaeuser at Texas A&M University. He spent more than 15 years researching social conditions at the Hachirogata reclaimed land area in Akita Prefecture, which culminated in the publication of *Ogata-Mura: Sowing Dissent and Reclaiming Identity in a Japanese Farming Village*, by Berghahn Books in 2012 (scheduled for release in paperback in October of 2015). He has also investigated tourism and the effects of depopulation in the Akita region, and was a contributor to the edited volume, *Japan's Shrinking Regions in the 21st Century* (Cambria, 2011). Recently he has been contributing articles to Kyoto Journal. He can be reached at wood@med.akita-u.ac.jp